Modern Biotechnology

A Glossary

Modern Biotechnology
A Glossary

Jayarama Reddy
Professor
St. Joseph's Post Graduate and Research Centre
Bangalore

I.K. International Publishing House Pvt. Ltd.
NEW DELHI • BANGALORE

Published by

I.K. International Publishing House Pvt. Ltd.
S-25, Green Park Extension
Uphaar Cinema Market
New Delhi–110 016 (India)
E-mail : info@ikinternational.com

ISBN 978-93-82332-75-6

Published by Krishan Makhijani for I.K. International Publishing House Pvt. Ltd., S-25, Green Park Extension, Uphaar Cinema Market, New Delhi–110 016 and Printed by Rekha Printers Pvt. Ltd., Okhla Industrial Area, Phase II, New Delhi–110 020.

Preface

Like everything else in this universe even education is evolving. There is constant transformation in its loom towards diverse subjects and disciplines. In the Vedic epoch and also later in the period of the Greek philosophers, the teachers and scientists marked no boundaries to the fields of enquiry. They had the fortitude and determination to disentangle the mysteries of all areas of science. Hence, they were just called thinkers, or rather *interdisciplinarians*. Then started the era of specialization, which has been dividing and subdividing science into a myriad of branches. This has resulted in the rise of specialists and super-specialists. Scientists soon narrowed down their spotlight of research to the tiniest areas of science. This is noticeably prominent in the field of medicine. These radical changes were due because of the quantitative approach towards scientific research and also due to the profundity of enquiry. At present we are working at the subatomic levels and we have the ability to craft nano-machines of topmost exactitude. In order to pull off these feats one has to have acquaintance of not one but many fields of science. This drift is what is making scientists *interdisciplinarians* once again like in the long-gone era. Pedagogy is also restructuring syllabi and curricula to suit the current scientific attitude, approach and trend. The designation and perception of the term biotechnology has also changed over the years since its inception in 1980. This branch which was interdisciplinary even from its inception has now become much more interdisciplinary and encompasses almost all areas of science. This is what I refer to as **Modern Biotechnology** and this glossary has resulted from my conviction that there is hardly another field of science as interdisciplinary as this one.

Biotechnology is technology based on biology, especially when used in agriculture, food science, and medicine. The UN Convention on Biological Diversity has come up with one of many definitions of biotechnology: "Biotechnology means any technological application that uses biological systems, living organisms, or derivatives thereof, to make or modify products or processes for specific use." This interdisciplinary branch covers many of the tools and techniques that are commonplace today in agriculture and food production. If interpreted in a narrow sense to consider only the "new" DNA, molecular biology and reproductive technology, the definition covers a range of different technologies, including gene manipulation, gene transfer, DNA typing and cloning of mammals. It is a general term used about a very broad field of study. According to the Convention on Biological Diversity, biotechnology means: "any technological application that uses biological systems, living organisms, or derivatives thereof, to make or modify products or processes for specific use." The swiftness of change in the sector means that terminology is constantly evolving, and yesterday's buzzword is today's jargon, and might well be tomorrow's mainstream term. The rate of evolution of terminology has been such that it has been very difficult to remain abreast of current usage. The idea for such a collection of terminology associated with the rapidly expanding fields related to or deriving from biotechnology and genetic engineering, was stimulated by the difficulty of communicating effectively in discussions at intergovernmental level. On various occasions, simple differences of interpretation of terminology have threatened to derail negotiations of international importance.

In the last two decades between 1990 and 2010 all areas of this discipline have been flourishing worldwide. There has been a tremendous increase in the number of biotech products in India. They accounted for about US$ 1789 million as far back as 1999. The same is expected to touch US$ 5000 million in 2013. Over the next five years, biotechnology can offer opportunities for fresh investment of Rs. 7 to 8 billion in India. This fresh investment could result in a turnover of Rs. 9 to 10 billion during the next 5 to 7 years. During 2005, the 329 publicly held biotechnology companies in the United States earned $47.8 billion in combined revenue, with market capitalization of $410 billion. Meanwhile, the 1086 private U.S. biotech firms earned revenues of $2.9 billion. Jobs for biological technicians, a key biotechnology occupation, are expected to grow by 28.2 per cent between 2004 and 2014, while those for biological scientists are projected to grow by 17.0 per cent. The biotechnology industry employed 7,13,000 workers in 2002 and five years later, in 2007, it employed 8,14,000. To succeed and grow in the 21st century economy, biotechnology employers need to fill each position in their companies — from entry-level to the most advanced — with qualified and skilled individuals. Because the industry is experiencing such rapid growth, biotechnology firms often demand more skilled workers than are available and are projected to need more workers than are currently enrolled in training programs. There is a tremendous potential to biotechnology to become the most sought after area of research in future. The need of the hour is to train the manpower to meet the demand. Though there are biotechnologists who are ready to take up tasks, we still are lacking in skilled professionals.

A glossary generally provides an excellent opportunity to collect and define terms that are either obscure or very local in nature. A glossary is an alphabetical list of terms peculiar to a field of knowledge with definitions or explanations. The Greek term, *gloss* refers to a brief notation of the meaning of a word or wording in a text. Terms in this volume have been alphabetized such that those beginning with numerals or Greek letters are inserted as if they were spelled out in Roman letters. This is the first edition of the ***Glossary of Modern Biotechnology***. I have attempted to be as comprehensive as possible, though some versions of terms (plurals, modified terms, unusual combinations, etc.) may not be so linked, nor are some common words that are not used in scientific manner as they are defined in the *Glossary*.

It is hoped that this glossary will become a ready reckoner for biotechnologists of today and tomorrow. It tries to provide a consolidated, comprehensive and yet accessible list of terms and acronyms that are used regularly in biotechnology *sensu lato* and in the very broad area commonly dubbed "genetic engineering," with all the associated problems of usage of originally discrete technical terms in a general context by the mass media that does not discriminate, or in a legal context that requires exact definitions. This glossary is an attempt to present an up-to-date list of terms currently in use in biotechnology, genetic engineering and allied fields. It is intended to provide a convenient reference source for researchers, students and technicians. The glossary should also be of particular value to those whose native language is not English. The terms included in the glossary have been selected by examination of books, dictionaries, journals and abstracts dealing entirely or in part with biotechnology or allied fields.

Management of biological information with the help of computer science is called bioinformatics. Computer science is used to gather, store, analyze and integrate biological and genetic information which can then be applied to gene-based drug discovery and development. The need for bioinformatics capabilities has been precipitated by the explosion of publicly available genomic information resulting from the unending number of genome projects. The interdisciplinary science of bioinformatics, a combination of molecular biology and computer science, is essential for the use of genomic information in understanding human diseases and in the identification of new molecular targets for drug discovery. In recognition of this, many universities, government institutions and pharmaceutical firms have formed bioinformatics groups, consisting of computational biologists and bioinformatics computer scientists. This glossary also includes important words related to bioinformatics.

Much of the information given in this book is a result my research in biotechnology, bioinformatics and biostatistics for the last 21 years. I am thankful to a number for people who have helped me in understanding biological sciences. The first name that I must take while thanking people who have helped me is that of Dr. R. M. Ranganath, Professor and former Registrar of Bangalore University, because it is in association with him that I have learnt a lot. He has always been my philosopher and guide and I thank him for standing behind me all the time and supporting me in all my academic endeavors. I am also extremely thankful to Dr. S. Sundara Rajan, the Dean of Research at Jain University. I would like to express gratitude to Dr. V. Niranjan, principal scientist and group leader at Genome Informatics at Genotypic Technology, who is the co-author for several research papers of mine. I thank Dr. Sitaram Annadana (Former Vice President of Avesthagen Ltd., Bangalore) through whom I had come in touch with Wageningen University, the Netherlands. St. Joseph's Institute of Gene Technology was started at St. Joseph's College, Bangalore in collaboration with Wageningen University, which started offering courses in bioinformatics and gene cloning and I was the coordinator of that institute. I take this opportunity to express my heartfelt gratitude and sincere thanks to Rev Dr. Daniel Fernandes, S. J., Principal of St. Joseph's College, Bangalore. I also would like to express my gratitude to Rev. Dr. Pravin Martis. S. J., the Rector of St. Joseph's College. I am deeply indebted to my colleagues, in the department of Botany at St. Joseph's College. I am also thankful to Dr. Nayeem Ullah, Khan, Dr. A.K.Bopaiah, Dr. Cheriyan Alexander, Dr. M.R.E. Wilson and Dr. T.V.Ranganathan for their help and encouragement. I owe my special thanks to my wife Sangeetha and my son Tarush for the timely help in all aspects of this project.

This attempt of mine, I presume, will lend a hand to the biologists in the hunt for specific details about the technical terms. I, at the same time welcome all proactive and constructive suggestions as I am positive that it will help in improving this Glossary when the next edition becomes due.

Jayarama Reddy

Contents

Dr. Jayarama Reddy is Professor at St. Joseph's Post Graduate Research Centre, Bangalore and has been teaching at graduate and postgraduate levels for the last twenty-three years. He is also the coordinator of the UGC recognized certificate course in bioinformatics at St. Joseph's College, Bangalore (www.sjc.ac.in). He obtained training in biochemical techniques from the Biochemistry laboratory at the Indian Institute of Science, Bangalore. For his Ph.D., he worked in the area of Biotechnology of Orchids with Dr. Devinder Prakash at the Orchid Laboratory, Department of Ornamental Crops, Indian Institute of Horticultural Research, Bangalore. As a part of his post-doctoral research he conducted investigations in gene cloning and bioinformatics at Centre for Life Technologies, Bangalore, a centre of Wageningen University, the Netherlands. Indira Gandhi National Open University honoured him with a gold medal for securing first rank at all India level in PGDHE. He has successfully completed several research projects in biotechnology and published nearly thirty-five research papers in national and international journals. He and his team worked on the cancer related protein, ***Mageel*** (melanoma antigen family E, 1) and submitted its structure to the Protein Data Bank. He published a monograph funded by the University Grants Commission on the medicinal plant, *Tribulus terrestris*. This is the fifth book by Dr. Reddy and he published his first book, " Biotechnology of Orchids" in 2008 (www.ikbooks.com). Dr. Reddy is recognized as the research guide by several academic institutions including Bangalore, Manipal, Bharatidasan, CMJ and Vinayaka Mission universities. Dr. Reddy has conducted three national conferences funded by the University Grants Commission in 2006, 2008, 2012 respectively at the capacity of Organizing Secretary and Chairman. He also served as the chairman of sessions during several national and international conferences. He is serving as the editor of the Registered National Journal of Plant Sciences, "Biovistas". Currently he is also the associate editor of International Journal of Bioinformatics Research (www.bioinfo.in). He is also the honorary director of Columbia Institutions, Bangalore (www.columbiainstitutions.org).

Characterization and statistical analysis of genes on human x-chromosome xq22.1, directory: X-chromosomal disorder, drug target for APG4A, comparative analysis of conserved sequences in swine flu infected individuals, **MUTVAC- tool to validate the vaccine candidate by simulation methods,** mutation analysis of DKC1 using structural bioinformatics approach, contig analysis of the X-chromosome, APG4A and its role in autophagy using bioinformatics approach, novel algorithm for predicting pseudogenes in human genes, designing a website GREPOR on RARA protein and studies on isoforms of human ubiquitin conjugating enzyme E2 are some the areas of bioinformatics in which the author has conducted research. He has designed and implemented a biological database (www.biologicaldatabank.com). Currently he is working on the application of biometry in the taxonomic identification of plant species and sequencing and analysis of genes of *Mycobacterium tuberculosis*.

A

AA amyloidosis: It is a form of amyloidosis associated with serum amyloid A protein, which is an acute-phase protein. It causes reactive systemic amyloidosis, as a "reaction" to inflammation.

AA: It is a permitted abbreviation for amino acid.

abiotic stress: It is the stress caused to the plants and animals by non-living factors which can harm the organisms.

abscisic acid (ABA): It was earlier called abscissin or dormin. It is a plant growth regulator involved in abscission, dormancy, stomatal opening/closure, and inhibition of seed germination. It also affects the regulation of somatic cell embryogenesis in some plant species.

abscisic acid antibody: It is a polyclonal antibody used against abscisic acid especially in the host, rabbit. It is actually synthetic 2-cis 4-trans-Abscisic acid conjugated to protein carriers.

absorb: It means to suck up or take in. In the cell, materials are taken in the form of solution.

absorption: It is the process of absorbing, like taking up of water and nutrients by assimilation or imbibition. It generally occurs by capillary, osmotic, chemical or solvent action, such as taking up of a gas by a solid or liquid, or taking up of a liquid by a solid.

absorption: In biology, absorption means the movement of fluids or dissolved substances into cells or across the tissues and organs through diffusion or osmosis.

abzyme: Abzymes are usually artificial constructs, but are also found in humans and in patients with autoimmune diseases such as systemic lupus erythematosus, where they can bind to and hydrolyze DNA. The term is derived from antibody and enzyme. It is also called *catmab* (from *catalytic monoclonal antibody*), which is a monoclonal antibody with catalytic activity. Molecules which are modified to gain new catalytic activity are called synzymes. Enzymes function by lowering the activation energy of the transition state, thereby catalyzing the formation of an otherwise less favourable molecular intermediate between reactants and products.

accession number: It refers to the identifier supplied by the curators of the major biological databases upon submission of a novel entry that uniquely identifies a sequence entry. RefSeq accession numbers are written in two letters followed by an underscore and six digits (e.g., NT_123456). The first two letters of the RefSeq accession number indicate the type of sequence included in the record. In library science, accession number refers to a sequential number given to each new book or a magazine.

acclimatization: It is the term often used in plant tissue culture. The ex-agar plants produced in vitro have to be transferred to in vivo conditions, during this period of transfer from culture environment to the outside conditions the plants undergo a process of adjustment process. In general, it is the adaptation of a living organisms (plant, animal or micro-organism) to a changed environment that subjects it to physiological stress.

accretion: It is the accumulation of marine sediments at the edges of a continent, building up in some cases into entire coastal mountain ranges. (See Plate Tectonics for more about what causes accretion).

acetyl co-enzyme-A ($C_{23}H_{38}N_7O_{17}P_3S$): A compound formed in the mitochondria when an acetyl group (CH3CO-) – derived from breakdown of fats, proteins, or carbohydrates – combines with the thiol group (-SH) of co-enzyme A.

acid rain: It is the precipitation heavy with nitric and sulfuric acid. Most of it is generated by sulfur dioxide and nitrogen dioxide (air pollution). Its pH is less than 5.6. Results include death of fish and plants, corrosion, groundwater pollution, and soil erosion. Its long-term effects are unknown.

acid: It is a substance with a pH less than 7 due to prevalent hydrogen ions. Acids tend to be sour and corrosive. The human stomach contains hydrochloric acid with a pH of 1; battery acid is stronger, but not by much. (Contrast with Basic).

acridine dyes: It is a class of positively charged polycyclic molecules that induces frame shift mutations by intercalating into DNA.

acridine orange: It is a nucleic acid selective fluorescent cationic dye useful for cell cycle determination. It is cell-permeable, and interacts with DNA and RNA by intercalation or electrostatic attractions respectively. When bound to DNA, it is very similar to fluorescein spectrally. Acridine orange will also enter acidic compartments such as lysosomes and become protonated and sequestered.

acridine yellow: It is a yellow dye with strong bluish-violet fluorescence. It is a derivative of acridine. In histology, it is used as a fluorescent stain and fluorescent probe for non-invasive measurements of cytoplasmic pH changes in whole cells. It is also used as a topical antiseptic. It can damage DNA and is used as a mutagen in microbiology.

acrylamide gel: It is also called polyacrylamide gel and is a separation matrix used in electrophoresis of biomolecules, such as proteins or DNA fragments. Traditional DNA sequencing techniques use these gels to separate DNA fragments differing by a single base-pair in length so that the sequence could be read. Most modern DNA separation methods now use agarose gels, except for particularly small DNA fragments.

actin: It is one of the two contractile proteins found in muscle. It is also found in the microfilaments that form part of the cytoskeleton of all cells.

activated charcoal (carbon): It is a form of charcoal which has been treated to remove hydrocarbons and increase its adsorptive properties. It acts by condensing and holding a gas or solute onto its surface, thus inhibitory substances in the nutrient medium may be adsorbed to charcoal included in the nutrient medium. Rooting factors such as phenolamines present as contaminants in charcoal may stimulate the growth *in vitro*. Its addition to rooting medium may stimulate root initiation in some plant species.

activator: A substance or physical agent that stimulates the transcription of a specific gene or operon. The term is also used for the compound that by binding to an allosteric site on an enzyme, enables the active site of the enzyme to bind to the substrate.

active site: A site on the surface of a catalyst at which enzymatic activity occurs. The term is used for the site on the surface of an enzyme molecule that binds the substrate molecule.

active site: It refers to the amino acid residues at the catalytic site of an enzyme. These residues provide the binding and activation energy needed to place the substrate into its transition state and bridge the energy barrier of the reaction undergoing catalysis.

adaptation: It how living things change, or what they do to survive in a particular environment. The organism is not a passive recipient of external circumstances; the relationship is interactive. (See Evolution).

adaptive (divergent) radiation: It is the evolution of new forms, sub-species or species from one species of plant or animal in order to exploit new habitats or food sources.

adaptive radiation: It refers to the evolution of many new species from a relative handful of ancestor species. It often happens after some kind of catastrophe empties a range of ecological niches simultaneously.

adaptor: It is a synthetic double-stranded oligonucleotide that has a blunt end, while the other end has a nucleotide extension that can base pair with a cohesive end created by cleavage of a DNA molecule with a specific type II restriction endonuclease. The blunt end of the adaptor can be ligated to the ends of a target DNA molecule and the construct can be cloned into a vector by using the cohesive ends of the adaptor.

addendum: It is the additional substance that is added as a nutrient during the preparation of tissue culture media. It is also known as adjuvant.

adenine (A): A constituent found in either DNA or RNA. It is a white crystalline purine base. It is also a constituent of nucleotides such as ADP and ATP. It is added to some tissue culture media, as adenine sulphate, to promote shoot formation and for its weak cytokinin effect.

adenosine disphosphate (ADP): A nucleotide consisting of adenine, sugar D-ribose, and two phosphate groups. ADP is formed by the removal of one phosphate from an ATP molecule by enzymes called ATPases. The formation of ADP from ATP releases usable metabolic energy. ADP is converted back to ATP by ATP synthases. Thus ADP is an important energy transfer molecule in cells.

adenosine monophosphate (AMP): Also known as 5'-adenylic acid, it is a nucleotide found in RNA. It is an ester of phosphoric acid and the nucleoside adenosine. AMP consists of a phosphate group, the sugar ribose, and the nitrogenous base, adenine.

adenosine triphospate (ATP): ATP is made from adenosine diphosphate (ADP) or adenosine monophosphate (AMP), and its use in metabolism converts it back into these precursors. ATP is therefore continuously recycled in organisms, with the human body turning over its own weight in ATP each day. It is used as a substrate in signal transduction pathways by kinases that phosphorylate proteins and lipids, as well as by adenylate cyclase, which uses ATP to produce the second messenger molecule cyclic AMP. It is called the energy currency of the cell and is a multifunctional nucleotide that plays an important role in cell biology as a coenzyme. ATP transports chemical energy within the cells for metabolism.

Adenovirus: It is a group of viruses that infect the membranes of the respiratory tract, the eyes, the intestines, and the urinary tract, which account for about 10% of acute respiratory infections in children and are a frequent cause of diarrhea.

ADEPT: (antibody-directed enzyme pro-drug therapy): It is a way to target a drug to a specific tissue. The drug is administered as an inactive pro-drug, and then converted into an active drug by an enzyme administered with a second injection. The enzyme is coupled to an antibody that concentrates it in the target tissue. When the enzyme arrives at the target tissue, the pro-drug is activated to form the active drug, while elsewhere it remains inactive.

adhesion: The mutual attraction between dissimilar molecules. It is also the phenomenon of sticking together of unlike substances, such as soil and water.

adiabatic cooling: It refers to the phenomenon when air masses expand and cool as they push up the side of a mountain.

A-DNA: It is a right-handed DNA double helix that has 11 base pairs per turn. DNA exists in this form when partially dehydrated.

adrenergic blocking agent: It inhibits response to sympathetic impulses by blocking the alpha or beta receptor sites of effector organs.

adsorbent: A substance to which compounds adhere. In tissue culture, an adsorbent is added to the culture medium to adsorb compounds released by cultured cells or tissues, thus minimizing any adverse effect on the subsequent growth in culture. A common adsorbent used in tissue culture is activated carbon or charcoal.

adsorption: It is the formation of a layer of gas, liquid or solid on the surface of a solid. Adsorption, the binding of molecules or particles to a surface, must be distinguished from absorption, the filling of pores in a solid. The binding to the surface is usually weak and reversible. Just about anything including the fluid that dissolves or suspends the material of interest is bound, but compounds with color and those that have taste or odor tend to bind strongly.

adult cloning: It can also be called organismal cloning. It is in fact the creation of identical copies of an adult animal by nuclear transfer from differentiated adult tissue. It should not be confused with gene cloning. Cloning of Dolly, the sheep, is the first example of this.

aeolian landform: It refers to sculpted by wind through deposition or erosion. Examples include dunes, deflation hollows, and sand-blasted outcroppings.

aeration (aerification): This refers to the process of supplying with or mix with air or gas. It is a process often used in large scale cell cultures in which the culture medium is continuously agitated by pumping the air.

aeration of soil: It refers to the extent of air gaps in soil. Aeration commonly refers to the process of using mechanized equipment to either puncture the soil with spikes.

aeration tank: A fluid-holding tank with provisions to aerate its contents by bubbling air or another gas through the liquid or by spraying the liquid into the air.

aerobe (aerobic organism): A microorganism that grows in the presence of oxygen. There are different types of aerobes. Obligate aerobes require oxygen for aerobic cellular respiration. In a process known as cellular respiration, these organisms use oxygen to oxidize substrates in order to obtain energy. Facultative anaerobes can use oxygen, but also have anaerobic methods of energy production. Microaerophiles are organisms that may use oxygen, but only at low concentrations.

aeroplankton: It refers to tiny organisms living in the atmosphere. Certain small seeds, bacteria, and spores are examples.

aerotolerant: It refers to the organisms that can survive in the presence of oxygen, but they are anaerobic because they do not use it as a terminal electron acceptor. It also refers to the ability of some bacteria to survive in the presence of oxygen, as in facultative anaerobes, which do not require oxygen for survival, but are not harmed by its presence.

affinity chromatography: A method for separating molecules by exploiting their ability to bind specifically to other molecules. There are several types of biological affinity chromatography. A biological molecule can be immobilized and a smaller molecule to which it is to bind can be stuck to it, or the smaller ligand can be immobilized and the macromolecule stuck to it.

affinity tag: The tag could be a short amino acid sequence, which is recognized by an antibody. An amino acid sequence that has been engineered into a protein to make its purification easier. These can work in a number of ways. The tag could also be a protein, which binds to some other material very tightly, thus allowing the protein to be purified by affinity chromatography.

aflatoxin: Toxic compounds, produced by moulds (fungi) of the *Aspergillus flavus* group, that bind to DNA and prevent replication and transcription. Aflatoxins can cause acute liver damage and cancer. Animals may be poisoned by eating stored food or feed contaminated with the mould. There are two main techniques that have been used most often to detect levels of aflatoxin in humans.

AFLP (amplified fragment length polymorphism): The AFLP technology has the capability to detect various polymorphisms in different genomic regions simultaneously. It is also highly sensitive and reproducible. This technique is divided into three steps: (1) Digestion of total cellular DNA with one or more restriction enzymes and ligation of restriction half-site specific adaptors to all restriction fragments. (2) Selective amplification of some of these fragments with two PCR primers that have corresponding adaptor and restriction-site specific sequences. (3) Electrophoretic separation and amplicons on a gel matrix, followed by visualisation of the band pattern.

agar: It is a polysaccharide and a solidifying agent used in nutrient solid media preparations. It is obtained from certain types of red algae. Both the type of agar and its concentration can affect the growth and appearance of cultured explants. It is critical that media have constant gel strength, as the effect on colony morphology can be dramatic.

agarose gel electrophoresis: A process in which a matrix composed of a highly purified form of agar is used to separate larger DNA and RNA molecules.

agarose: The main constituent of agar. Agarose is a polysaccharide obtained from agar that is used in a variety of life science applications especially gel electrophoresis. Agarose forms an inert matrix utilized in separation techniques. Many structures easily affix to agarose including various types of proteins.

agent orange: A herbicide containing 2,4,5-T and 2,4-D and the contaminant dioxin; it is suspected of being carcinogenic and teratogenic.

agents: It refers to the independent, autonomous, software modules that can search the Internet for data or content pertinent to a particular application, such as a gene, protein, or biological system.

aggradation: It refers to a downward accumulation of stream-carried inorganic matter. Often has the effect of making the bed of a stream or flood plain rise. Also, a phase of forest biomass accumulation in the years that follow a harvest.

agonist: A drug, hormone or transmitter substance that forms a complex with a receptor site that is capable of triggering an active response from a cell.

agribusiness: It refers to one that markets farm products and equipment, including warehousing, seed monopolization, and fertilizer. The corporatization of farming, resulting in a handful of very large non-local companies owning and managing—and in some cases ruining millions of high-yield acres.

***agrobacterium tumefaciens*:** A bacterium that causes crown gall disease in some plants. The bacterium infects a wound, and injects a short stretch of DNA into some of the cells around the wound. The DNA comes from a large plasmid—the Ti (tumour induction) plasmid—whose short region called T-DNA (transferred DNA) is transferred to the plant cell, where it causes the cell to grow into a tumour-like structure. The T-DNA contains genes which *inter alia* allow the infected plant cells to make two unusual compounds, nopaline and octopine, that are characteristic of transformed cells. The cells form a gall, which hosts the bacterium. This DNA-transfer mechanism is exploited in the genetic engineering of plants. The Ti plasmid is modified so that a foreign gene is transferred into the plant cell along with or instead of the nopaline synthesis genes. When the bacterium is cultured with isolated plant cells or with wounded plant tissues, the "new" gene is injected into the cells and ends up integrated into the chromosomes of the plant.

***agrobacterium tumefaciens*-mediated transformation:** A naturally occurring process of DNA transfer from the bacterium *A. tumefaciens* to plants. It is one of the most commonly used methods of genetic transformation in plant biotechnology.

agrobacterium: A genus of bacteria that includes several plant pathogenic species, causing tumour-like symptoms.

agrobiodiversity: Agrobiodiversity is the product of natural selection processes and the careful selection and inventive developments of farmers, herders and fishermen over millennia. Agrobiodiversity is a vital sub-set of biodiversity. Many people's food and livelihood security depend on the sustained management of various biological resources that are important for food and agriculture.

AIDS (acquired immunodeficiency syndrome): The fatal human disease in which the immune system is destroyed by a retrovirus (Human Immunodeficiency Virus, HIV). The virus infects and destroys helper T-cells, which are essential for combating infections. The symptoms of AIDS are primarily the result of conditions that do not normally develop in individuals with healthy immune systems. Most of these conditions are infections caused by bacteria, viruses, fungi and parasites that are normally controlled by the elements of the immune system that HIV damages. Opportunistic infections are common in people with AIDS. HIV affects nearly every organ system.

airlift fermenter: A cylindrical fermentation vessel in which the cells are mixed by air introduced at the base of the vessel and that rises through the column of culture medium. The cell suspension circulates around the column as a consequence of the gradient of air bubbles in different parts of the reactor.

airshed: It refers to an area characterized by air with common qualities. (Compare Watershed).

albedo: It refers to the luminosity shining from a reflective surface; earthshine is one type. About 1/3 of the sun's radiation is reflected back into space, with the remaining 42% warming the land and air and 23% moving water through the hydrologic cycle. (See Emissivity).

albinism: Hereditary absence of pigment in an organism. Albino animals have no colour in their skin, hair and eyes. The term is also used for the absence of chlorophyll in plants. An organism lacking pigmentation, due to genetic factors, is called an **albino**.

algal biomass: Single-celled plants, such as *Chlorella* spp. And *Spirulina* spp., are grown commercially in ponds to make feed material. Chlorella is grown commercially to make into fish food: it is fed to zooplankton, and these in turn are harvested as feed for fish farms. This is a means of converting sunlight into food in a way more convenient and controllable than normal farming.

algal bloom: It refers to explosion of a phytoplankton population, sometimes because of incoming pollutants that artificially enrich the waters with nutrients.

alginate: It refers to a gelatinous extract of brown seaweed used to make a number of products, including toothpaste, beer, various medicines, paper, and frozen foods.

alginate: Polysaccharide gelling agent. Alginate absorbs water quickly, which makes it useful as an additive in dehydrated products such as slimming aids, and in the manufacture of paper and textiles. It is also used for waterproofing and fireproofing fabrics, as a gelling agent, for thickening drinks, ice cream and cosmetics, and as a detoxifier that can absorb poisonous metals from the blood.

algorithm: It refers to a series of steps defining a procedure or formula for solving a problem that can be coded into a programming language and executed. Bioinformatics algorithms are typically used to process, store, analyze, visualize and make predictions from biological data.

alignment: It refers to a result of a comparison of two or more gene or protein sequences in order to determine their degree of base or amino acid similarity. Sequence alignments are used to determine the similarity, homology, function or other degree of relatedness between two or more genes or gene products.

alkaline: It refers to the pH over 7; emits hydroxyl ions. Also called "basic." It neutralizes the acids it combines with chemically.

alkylating agents: Chemicals that transfer alkyl (methyl, ethyl, etc.) groups to the bases in DNA.

allele (allelomorph): One of a pair, or series, of variant forms of a gene that occur at a given locus in a chromosome, and occupy the same relative position (locus) on homologous chromosomes. Within a population, there may be more than two alleles of a gene.

allele frequency: The number of copies of an allele in a population, expressed as a proportion of the total number of copies of all alleles at a locus in a population is called allelic frequency.

allele-specific amplification (ASA): The use of polymerase chain reaction (PCR) at a sufficiently high stringency so that only a primer with exactly the same sequence as the target DNA will be amplified. It is a powerful means of genotyping single-locus disorders that have been characterized at the molecular level.

allelic exclusion: A phenomenon whereby only one functional allele of an antibody gene can be assembled in a given B lymphocyte. The "allele" on the other homologous chromosome in a diploid mammalian cell cannot undergo a functional rearrangement, which would result in the production of two different antibodies by a single plasma cell.

allelopathy: It refers to the metabolic impact of one plant on another, whether beneficial or harmful. For example, eucalyptus tree toxins that inhibit the growth of certain plants.

allelopathy: The phenomenon by which the secretion of chemicals, such as phenolic and terpenoid compounds, by a plant inhibits the growth or reproduction of other plant species with which it is competing.

allochthonous: It refers to something organic imported into an ecosystem from outside of it.

allogenic: It refers to originating outside a system.

allopatric: It refers to separated, apart, as with a group of organisms living away from the original group.

allosteric regulation (control): A catalysis-regulating process in which the binding of a small effector molecule to one site on an enzyme affects the activity at another site. Reversible interaction of a small molecule with a protein molecule, resulting in a change in the shape of the protein and consequent alteration of the interaction of that protein with a third molecule is called **allosteric transition**.

allozymes: They are the enzymes from different alleles of the same gene, and isozymes represent enzymes from different genes that process or catalyse the same reaction; the two words are usually used interchangeably.

alluvium: It refers to the sediment transported by water, usually river or stream water.

alpha diversity (local diversity): It refers to the variety of organisms in a given habitat or location.

alpha-lactalbumin: One of the golden ingredients that makes a mother's milk so special is Alpha-lactalbumin, also known as a protein. There are two categories of milk protein in the breast milk—whey and casein proteins. Unlike cow's milk that has mostly casein protein (20% whey vs 80% casein), human milk has more whey than casein (60% whey vs 40% casein). Alpha-lactalbumin (alpha-protein) is the major whey protein found in breast milk.

alpine glacier (mountain glacier): It refers to a small glacier sitting in a U-shaped mountain valley.

alpine: It refers to above the timberline. Roughly synonymous with "mountainous."

alternative fuels: It refers to fuels from sources cleaner than coal or petroleum products: ethanol, methanol, natural gas, solar, wind, geothermal, biodiesel from vegetable oil, etc.

alternative mRNA splicing: The inclusion or exclusion of different exons to form different mRNA transcripts.

alternative splice-form: It refers to one of the possible alternate combinations of exons into a folded protein that are possible by recombining multiple gene segments during mRNA splicing in higher organisms.

alternative splicing: It refers to one of the alternate combinations of a folded protein that are possible due to recombination of multiple gene segments during mRNA splicing that occurs in higher organisms.

alterne: It refers to the plant communities that alternately occupy a territory.

Alu family: It refers to the common set of dispersed DNA sequences found throughout the human genome; each is about 300 bases long and they are repeated at least 500,000 times. Alu sequences are speculated to have originated from viral RNA sequences that integrated into human DNA thousands of years ago.

alu sequences: A family of 300-bp sequences occurring nearly a million times in the human genome.

ambient temperature: Air temperature at a given time and place. The temperature of the medium surrounding an object. The term is often used to denote prevailing room temperature.

ambient: It refers to the prevailing natural conditions studied and recorded outside rather than indoors under microscopes or other controlled conditions.

amendment: It refers to something added to fix chemically troubled soil. Potash added to soil poor in potassium, for example.

amensalism: It refers to a one-sidedly harmful relationship between dissimilar organisms.

amino acid: An acid containing the group NH^2. In particular, any of 20 basic building blocks of proteins with a free amino (NH_2) and a free carboxyl (COOH) group, and having the basic formula NH_2 – CR – COOH. According to the side group R, they are subdivided into: polar or hydrophilic (serine, threonine, tyrosine, asparagine and glutamine); nonpolar or hydrophobic (glycine, alanine, valine, leucine, isoleucine, proline, phenylalanine, tryptophan and cysteine); acidic (aspartic acid and glutamic acid) and basics (lysine, arginine, hystidine). The sequence of amino acids determines the shape, properties and the biological role of a protein. Plants and many micro-organisms can synthesize amino acids from simple inorganic compounds, but animals are unable to synthesize some of them, called essential amino acids, so they must be present in the diet. (*Also see the appendix-1*).

aminoacyl site (A-site): One of the two sites on ribosomes to which the incoming aminoacyl tRNA binds.

aminoacyl tRNA synthetase: Enzyme that attaches each amino acid to its specific tRNA molecule.

amniocentesis: A procedure for obtaining amniotic fluid from a pregnant mammal for diagnosis of some diseases in the unborn foetus. Cells are cultured, and metaphase chromosomes are examined for irregularities (e.g., Down syndrome, spina bifida, etc., in humans).

Amoeba (*Dictyostelium discoideum*): It is a powerful model organism because of its rapid generation time, facile reverse genetics and its ease of use in the study of many areas of cell and molecular biology. *Dictyostelium* cells grow as unicellular organisms but, upon starvation, aggregate to form a multicellular tissue capable of differentiating into multiple cell types. During this developmental program, the amoebae employ many cellular processes to ultimately form a spore-bearing fruiting body, including chemotaxis, cytoskeletal rearrangement, signal transduction, cell sorting, and pattern formation. Because the life cycle consists of both unicellular and multicellular phases, *Dictyostelium* provides insight into processes needed for multicellularity.

amorph mutation (null mutation): A mutation that obliterates gene function.

amperometric titration: It refers to a class of titrations in which the equivalence point is determined through measurement of the electric current produced by the titration reaction. It is a form of quantitative analysis.

amphimixis: True sexual reproduction involving the fusion of male and female gametes and the formation of a zygote.

ampicillin: A penicillin-derived antibiotic that prevents bacterial growth by interfering with the synthesis of cell wall.

amplification (gene): A treatment designed to increase the proportion of plasmid DNA relative to that of bacterial (host) DNA. It is also the process of replication of a gene library in bulk or the creation of many copies of a segment of DNA by polymerase chain reaction (PCR).

amylase: A group of enzymes that degrades starch, glycogen and other polysaccharides, producing a mixture of glucose and maltose.

amylase: A polysaccharide consisting of linear chains of 100 to 1000 glucose molecules. The water-soluble portion of starch.

amyloidosis: It refers to a variety of conditions in which amyloid proteins are abnormally deposited in organs and/or tissues. A protein is described as being amyloid if, due to an alteration in its secondary structure, it takes on a particular aggregated insoluble form similar to the beta-pleated sheet.

amylolysis: The process of breaking down starch into sugars.

amylopectin: A polysaccharide comprising highly branched chains of glucose molecules. The water-insoluble portion of starch.

anabolic pathway: A biosynthetic pathway by which a metabolite is synthesized.

anaerobic respiration: Respiration in which the foodstuffs are partially oxidized, with the release of chemical energy, in a process not involving atmospheric oxygen. For example, alcoholic fermentation, in which one of the end products is ethanol.

anaerobic: An environment or condition in which molecular oxygen is not available for chemical, physical or biological processes.

anagensis: It refers to evolutionary change, but without spilling over into speciation.

analogue forestry: It refers to a method for restoring ecosystems, developed from local Sri Lankan home gardens by the Neo-Synthesis Research Centre (NSRC), that seeks to bring back what grew there originally.

analogy: It refers to the reasoning by which the function of a novel gene or protein sequence may be deduced from comparisons with other gene or protein sequences of known function. Identifying analogous or homologous genes via similarity searching and alignment is one of the chief uses of Bioinformatics.

Analysis of variance (ANOVA): It is a statistical technique used to test the equality of three or more sample means and thus make inferences as to whether the samples comes from populations having the same mean.

anapsid: It refers to a vertebrate whose skull contains no side openings behind the eyes. The only living examples are turtles.

anchor gene: A gene that has been positioned on both the physical map and the linkage map of a chromosome.

androgen: Any hormone that stimulates the development of male secondary sexual characteristics, and contributes to the control of sexual activity in vertebrate animals. Usually synthesized by the testes.

androgenesis: Male parthenogenesis, i.e., the development of a haploid embryo from a male nucleus. The maternal nucleus is eliminated or inactivated subsequent to fertilization of the ovum, and the haploid individual contains in its cells only the genome of the male gamete. Androgenesis is detected by cytological staining.

aneuploid: An organism or cell having a chromosome number that is not an exact multiple of the monoploid *(x)* with one chromosome being present in greater (e.g., trisomic $2n + 1$) or lesser (e.g., monosomic $2n - 1$) number than the normal diploid number.

angle of repose: It refers to the steepest angle that slope, rock, or detritus material settles into without toppling. Builders of flood-prone roads and sliding hillside homes in California ignore the angle of repose so often that it could be renamed the angle of depose.

animal cell immobilization: Entrapment of animal cells in some solid material in order to produce some natural product or genetically engineered protein. Animal cells have the advantage that they already produce many proteins of pharmacological interest, and genetically engineered proteins are produced by them with post-translation modifications normal to animals. However, since animal cells are much more fragile than bacterial cells, they cannot tolerate a commercial fermentation process. So typical materials used are hollow fibre membrane bioreactors, or porous carriers made of polysaccharide, protein, plastic or ceramic materials with microscopic holes inside which the cells grow.

animism: It refers to a derogatory anthropological term for what most human cultures have believed throughout prehistory: that the earth is alive and reactive, as are its many places. Greeks and Romans once thought a "genius loci" or spirit of place inhabited every hill, grove, and stream. Such beings still live in all human mythologies. The modern counterpart is panpsychism, the idea that all things possess qualities of mindfulness or psyche. With the coming of heavy industry, such ideas gave way to the financially convenient reduction of the earth to the status of a lifeless resource.

annealing: The process of heating (denaturing step) and slowly cooling (renaturing step) double-stranded DNA to allow the formation of hybrid DNA or complementary strands of DNA or DNA and RNA. The sticking together (renaturing) of complementary single strands of DNA to make a double-stranded DNA after the strands have first been pulled apart (denatured). This involves forming hydrogen bonds between the base pairs.

annotation: It refers to the combination of comments, notations, references, and citations, either in free format or utilizing a controlled vocabulary, that together describe all the experimental and inferred information about a gene or protein. Annotations can also be applied to the description of other biological systems. Batch, automated annotation of bulk biological sequence is one of the key uses of bioinformatics tools.

anonymous DNA marker: A DNA marker detectable by virtue of variation in its sequence, irrespective of whether or not it actually occurs in or near a coding sequence. Microsatellites are typical anonymous DNA markers.

anorexia: From an ecopsychological perspective, it refers to a horror of carnality so profound that it seeks to etherealize the body itself. Tends to be over represented in highly industrialized nations.

antemortem: It refers to "before slaughter." The USDA meat inspectors conduct animal examination shortly before they are butchered.

anther culture: It is one of the very popular methods for the production of haploids through the culture anthers or microspores on artificial culture medium. This leads to the growth of microspores into plantlets *in vitro*. The technique was developed in 1964 by Guha and Maheshwari.

anthocyanin: Water-soluble blue, purple and red flavonoid pigments found in the vacuoles of cells.

anthropocentrism: It refers to human chauvinism, according to John Seed. An example is the belief that the earth is merely a stage for human salvation or self-development without any intrinsic importance of its own.

antiauxin: A chemical that interferes with the auxin response. Antiauxin may or may not involve the prevention of auxin transport or movement in plants. Some antiauxins are said to promote morphogenesis *in vitro,* such as 2,3,5-tri-iodobenzoate (TIBA), or 2,4,5-trichlorophenoxyacetate (2,4,5-T), which stimulate the growth of some cultures.

antibiosis: The prevention of growth or development of an organism by a substance or another organism.

antibiotic (bactericide or bacteriostat): A class of natural and synthetic compounds that inhibit the growth of or kill some microorganisms. Antibiotics such as penicillin are often used to control (to some extent kill) contaminating organisms. However, resistance to particular antibiotics can be acquired through mutations. Some contaminating organisms are only suppressed or their metabolism slowed to an insignificant level.

antibiotic resistance: It refers to the ability of a microorganism to produce a protein that disables an antibiotic or prevents the transport of the antibiotic into the cell.

antibody: An immunological protein (also called immunoglobulin or Ig) produced by certain white blood cells (lymphocytes) of the immune system of an organism in response to a contact with a foreign substance (antigen). It has the ability of specifically binding with the foreign substance and rendering it harmless. Depending on the type of heavy chain present, there are five classes of antibodies: IgA, IgD, IgE, IgG, and IgM. Antibodies have a well-defined structure. Each antibody has two identical "light" chains and two identical "heavy" chains. Each chain comprises a constant region, i.e., a region that is the same between antibodies of the same class and sub-class, and a variable region that differs between. The antigen-binding region or binding site – *complementarity determining region* – is in the variable region. The antibody can be cut by proteases into several fragments, known as Fab, Fab', and Fc. The synthesis of antibodies by B cells in response to an encounter of the cells of the immune system with a foreign immunogen is called antibody-mediated (humoral) immune response.

anticoding strand: The strand of the DNA double helix that is actually transcribed; also known as the antisense or template strand.

anticodon: A triplet of nucleotides in a tRNA molecule that pairs with a complementary triplet of nucleotides, or codon, in an mRNA molecule during translation. *See* codon; mRNA; RNA; tRNA.

antigen (immunogen): A compound that elicits an immune response by stimulating the production of antibodies. The antigen, usually a protein, when introduced into a vertebrate organism is bound by the antibody or a T cell receptor.

antigenic determinant (epitope): A surface feature of a microorganism or macromolecule, such as a glycoprotein, that elicits an immune response.

antigenic switching: The altering of a microorganism's surface antigens through genetic re-arrangement to elude detection by the host's immune system.

antihaemophilic globulin: Blood globulin that reduces the clotting time of haemophilic blood.

anti-idiotype antibodies: Antibodies which recognize the binding sites of other antibodies. Their binding sites are complementary to the binding sites of another immunoglobulin. When an animal becomes immune to something, it not only acquires antibodies against that something, but also acquires antibodies against those antibodies. This forms a network of antibodies which can all bind to each other to various degrees, helping to regulate the immune response. Some allergic responses are in part due to the breakdown of this sort of regulation.

antinutrients: Compounds that inhibit the normal uptake of nutrients.

anti-oncogene: A gene whose product prevents the normal growth of tissue.

antioxidant solution: Pre-treatment solution (e.g., vitamin C, citric acid) that retards the senescence and browning of tissue. It is employed to incubate explants prior to surface sterilization.

antioxidant: Compound that slows the rate of oxidation reactions.

antiparallel orientation: The normal arrangement of two strands of a DNA molecule, and other nucleic-acid duplexes (DNA-RNA, RNA-RNA), in which the two strands are oriented in opposite directions so that the 5′-phosphate end of one strand is aligned with the 3′-hydroxyl end of the complementary strand.

antisense DNA: The strand of chromosomal DNA that is transcribed or the DNA sequence that is complementary to all or part of an mRNA molecule.

antisense gene: A gene that produces a transcript (mRNA) that is complementary to the pre-mRNA or mRNA of a normal gene (usually constructed by inverting the coding region relative to the promoter).

antisense RNA: An RNA sequence that is complementary to all or part of a functional mRNA molecule, to which it binds, blocking its translation.

antisense therapy: The *in vivo* treatment of a genetic disease by blocking the translation of a protein with a DNA or an RNA sequence that is complementary to a specific mRNA.

antiseptic: Any substance that kills or inhibits the growth of disease causing microorganism (a microorganism capable of causing sepsis), but is essentially non-toxic to cells of the body.

antiserum: The fluid portion of the blood of an animal (after coagulation of the blood), containing antibodies.

anti-terminator: A type of protein which enables RNA polymerase to ignore certain transcriptional stop or termination signals and read through them to produce longer mRNA transcripts.

apical dominance: The phenomenon of inhibition of growth of lateral (axillary) buds in a plant by the presence of terminal (apical) bud on the branch, due to auxins produced by the apical bud.

apical meristem: A region of the tip of each shoot and root of a plant in which cell division is continually occurring to produce new stem and root tissue, respectively.

apoenzyme: Inactive enzyme that has to be associated with a specific organic molecule called a co-enzyme in order to function. The apoenzyme/co-enzyme complex is called a holoenzyme.

apoptosis: The process of cell death, which occurs naturally as a part of normal development, maintenance and renewal of tissue in an organism. It is different from necrosis which means cell death caused by a toxic substance.

aquaculture: It refers to growing and harvesting fish and shellfish in land-based ponds. Relative protein yields often exceed those of land cropping by 4-20 times. Ponds attract beneficial wildlife, cool the surrounding areas, reflect sunlight, draw birds, and make convenient places for growing rice and other moisture-loving plants. (See also Hydroponics).

***Arabidopsis thaliana*:** It is a small flowering plant that belongs to the Brassica family, which includes species such as broccoli, cauliflower, cabbage, and radish. Because *Arabidopsis* has a small genome relative to other plants and is easily grown under laboratory conditions, it has become the organism of choice for basic studies of molecular genetics of flowering plants. Scientists expect that systematic studies of *Arabidopsis* will offer important advantages for basic research in genetics and molecular biology and will illuminate numerous features of plant biology, including those of significant value to agriculture, energy, environment, and human health.

***Arabidopsis*:** A genus of flowering plants in the Cruciferae. *A. thaliana* is used in research as a model plant because it has a small genome (5 pairs of chromosomes; 2n = 10) and can be cultured easily, with a generation time of two months.

archaea: It is a group of some of the most fascinating organisms on the earth. Historically, they have been considered part of the bacteria, but recent work has shown them to be very distinct. Archaea are unique for two reasons. First, they represent the most primitive organisms yet discovered and have been found to have a remarkable resemblance to some of the earliest known fossils. Second, they live in some of the most extreme environments on the earth, hence the nickname "extremophiles". Research studies conducted on the molecules that enable extremophiles to prosper are fast becoming useful in areas such as bioremediation, global climate change, biotechnology, and energy production. In addition, although archaeans resemble bacteria and have some genes that are similar to bacterial genes, they also have other genes that resemble eukaryotes and some genes that appear to be unique. Studies of these unshared genes may provide new clues to the evolution of early life on earth.

ARS (autonomous replicating sequence): Any eukaryotic DNA sequence that initiates and supports chromosomal replication; they have been isolated in yeast cells. Also called autonomous(ly) replicating segment.

artificial in-embryonation: Non-surgical transfer of embryo(s) to a recipient female. As *in vitro* embryo technology develops, artificial in-embryonation will gradually replace artificial insemination.

artificial insemination: The deposition of semen, using a syringe, at the mouth of the uterus to make conception possible. It is used in the breeding of domestic animals.

artificial neuron: It is a mathematical function conceived as a crude model or abstraction of biological neurons. Artificial neurons are the constitutive units in an artificial neural network. Depending on the specific model used, it can receive different names, such as semi-linear unit, Nv neuron, binary neuron, linear threshold function or McCulloch–Pitts neuron. The artificial neuron receives one or more inputs (representing the one or more dendrites) and sums them to produce an output (synapse). Usually the sums of each node are weighted, and the sum is passed through a non-linear function known as an activation function or transfer function. The transfer functions usually have a sigmoid shape, but they may also take the form of other non-linear functions, piecewise linear functions, or step functions. They are also often monotonically increasing, continuous, differentiable and bounded. Artificial selection (selective breeding). It describes intentional breeding for certain traits, or combination of traits. The term was utilized by Charles Darwin in contrast to natural selection, in which the differential reproduction of organisms with certain traits is attributed to improved survival or reproductive ability ("Darwinian fitness"). As opposed to artificial selection, in which humans favor specific traits, in natural selection the environment acts as a sieve through which only certain variations can pass.

artificial nutrient medium (culture medium): The complex mixture of defined organic and inorganic nutrients for growing cells, tissues, organs, plants, bacteria or any other organisms is called a culture medium. Murashige and Skoog (MS) medium is one of the most widely used media in plant tissue culture.

artificial seed (synthetic seeds): Encapsulated or coated somatic embryos (embryoids) that are produced *in vitro*, planted and treated like seed.

ascorbic acid (vitamin C): It is a water-soluble vitamin present naturally in some plants; it is also produced synthetically. Aside from its role as a vitamin, it is used as an antioxidant in plant tissue culture; and included in disinfection solutions. It is a sugar acid with antioxidant properties.

a-site (aminoacyl site): The site on the ribosome occupied by an aminoacyl- tRNA just prior to peptide bond formation.

assay: It refers to the method for measuring a biological activity. This may be enzyme activity, binding affinity, or protein turnover. Most assays utilize a measurable parameter such as color, fluorescence or radioactivity to correlate with the biological activity.

assembly: It refers to the compilation of overlapping sequences from one or more related genes that have been clustered together based on their degree of sequence identity or similarity. Sequence assembly may be used to piece together "shotgun" sequencing fragments (see shotgun sequencing) based upon overlapping restriction enzyme digests, or may be used to identify and index novel genes from "single-pass" cDNA sequencing efforts.

assortative mating: Mating in which the partners are chosen on the basis of phenotypic similarity.

asynapsis: The failure or partial failure in the pairing of homologous chromosomes during the meiotic prophase.

ATP (adenosine triphosphate): A nucleotide of fundamental importance as a carrier of chemical energy in all living organisms. It consists of adenosine with three phosphate groups (adenosine) through its ribose (sugar) portion. Upon hydrolysis, bonds yield either one molecule of ADP (adenosine diphosphate) and an inorganic phosphate, or one molecule of AMP (adenosine monophosphate) and pyrophosphate; in both cases releasing energy that is used to power biological processes. ATP is regenerated by rephosphorylation of AMP and ADP, using chemical energy derived from the oxidation of food.

ATP-ase: An enzyme that brings about the hydrolysis of ATP, by the cleavage of either one phosphate group with the formation of ADP and inorganic phosphate, or of two phosphate groups, with the formation of AMP and pyrophosphate.

attenuated vaccine: A virulent organism that has been modified to produce a less virulent form, but nevertheless retains the ability to elicit antibodies against the virulent form.

attenuation: A mechanism for controlling gene expression in prokaryotes that involves premature termination of transcription. Attenuator is a nucleotide sequence in the 5′ region of a prokaryotic gene (or in its RNA) that causes premature termination of transcription, possibly by forming a secondary structure.

authentic protein: A recombinant protein that has all the properties—including any post-translational modifications—of its naturally occurring counterpart.

autocatalysis: Catalysis in which one of the products of the reaction is a catalyst for the reaction. Usually, the catalysis starts slowly and increases as the quantity of the catalyst increases, falling off as the product is used up.

autoclave: An enclosed chamber in which substances are heated under pressure to sterilize utensils, liquids, glassware, etc., using steam. It is used to sterilize equipment and supplies by subjecting them to high pressure saturated steam at 121°C for around 15–20 minutes depending on the size of the load and the contents.

auto-immunity: A disorder in the body's defence mechanism in which an immune response is elicited against its own (self) tissues. Disorders in which the immune systems of affected individuals produce antibodies against molecules that are normally produced by those individuals (called self-antigens) are called **autoimmune diseases.** Rheumatoid arthritis (RA), Lupus (Systemic Lupus Erythematosus), Goodpasture's syndrome and Polymyalgia Rheumatica are some of the common autoimmune diseases.

autologous cells: Cells that are taken from an individual, cultured (or stored), and, possibly, genetically manipulated before being infused back into the original donor.

autolysis: The process of self-destruction of a cell, cell organelle, or tissue. It occurs by the action of lysosomic enzymes.

autoradiography: A technique that captures the image formed in a photographic emulsion as a result of the emission of either light or radioactivity from a labelled component that is placed next to unexposed film. The technique is used for detecting the location of an isotope in a tissue, cell or molecule. The sample is placed in contact with a photographic emulsion, usually an X-ray film. The emission of Ü-particles from the sample activates the silver halide grains in the emulsion and allows them to reduce to metallic silver when the film is developed. In genetic engineering, autoradiography is most commonly used to detect the hybridization of a radioactive DNA (probe) molecule to denatured DNA in either the Southern transfer or colony hybridization procedures. A picture prepared by labelling a substance such as DNA with a radioactive material such as tritiated thymidine and allowing the image produced by decay radiation to develop on a film over a period of time is called an **autoradiograph.**

autoradiography: It refers to the method used to locate radioisotope-labeled materials which have been separated in gels or are present in blots. The location of the radiolabeled material is determined by overlaying the test material with a photographic film that is sensitive to the radioisotope.

autotrophic: Self-nourishing organisms capable of utilizing carbon dioxide or carbonates as the sole source of carbon and obtaining energy for life processes from radiant energy or the oxidation of inorganic elements, or compounds such as iron, sulphur, hydrogen, ammonium and nitrites.

autotrophy: It is the capacity of an organism to use light as the sole energy source in the synthesis of organic material from inorganic elements or compounds. Autotrophic organisms include green photosynthesizing plants and some photosynthetic bacteria.

auxin: A group of plant growth regulators (natural or synthetic) which stimulate cell division, enlargement, apical dominance, root initiation, and flowering. One naturally produced auxin is indole-acetic acid (IAA).

auxin-cytokinin ratio: The relative proportion of auxin to cytokinin present in plant-tissue-culture media. Varying the relative amounts of these two hormone groups in tissue culture formulae affects the proportional growth of shoots and roots *in vitro*. As the ratio is increased (increased auxin or decreased cytokinin), roots are more likely to be produced, and as it is decreased, root growth declines and shoot initiation and growth are promoted. This relationship was first recognized by Miller and Skoog.

auxotroph: A mutant cell or microorganism lacking the capacity to form an enzyme or metabolite present in the parental strain, and that consequently will not grow on a minimal medium, but requires addition of some compounds such as an amino acid or a vitamin for growth.

average: It is a single value that describes the characteristics of the entire mass of data.

axenic culture: A culture free of external contaminants and internal symbionts. This is generally not possible with surface sterilization alone.

azimuth: It refers to the angular distance clockwise along the horizon between the north and the position of a celestial object. A star hanging exactly over the northern point on the horizon has an azimuth of 0°, for example, and one located exactly east an azimuth of 90°. Azimuth is combined with altitude (the distance of an object above the horizon) to calculate the direction of an object as seen from a specific earthly location.

B

B cells (B lymphocytes): An important class of white blood cells that mature in bone marrow and produce antibodies. They are largely responsible for the antibody-mediated or humoral immune response. They give rise to antibody-producing plasma cells and some other cells of the immune system.

B lymphocytes (B cells): The abbreviation "B" comes from the bursa of Fabricius in birds, where they mature. In mammals, immature B cells are formed in the bone marrow. These are an important class of lymphocytes that mature in bone marrow and the Bursa of Fabricius, which are largely responsible for the antibody-mediated or humoral immune response. They give rise to antibody producing plasma cells and some other cells of the immune system.

BAC (bacterial artificial chromosome): A cloning vector constructed from bacterial fertility (F) factors; like YAC vectors, they accept large inserts of size 200 to 500 kb.

***Bacillus thuringiensis* (Bt):** A bacterium that kills insects as it contains a gene called Bt gene. Bt gene is being used extensively in plant biotechnology in order to produce varieties of plants resistant against insect larvae of certain genera. It is also a major component of the microbial pesticide industry.

Bacillus: A rod-shaped bacterium.

background extinction rate: It refers to the natural rate of extinction for a species. It is in contrast sharply with mass extinction.

backscattering: It refers to solar radiation reflected back into the space by particles in the atmosphere.

backshore: It refers to the strip of beach above the daily tides, but within the reach of storm waves.

bacterial artificial chromosome (BAC): It is a cloning vector that can incorporate large fragments of DNA. (See also YACs.)

bacteriocide (bactericide): A chemical or drug that kills bacterial cells.

bacteriocin: A protein produced by the bacteria of one strain that is active against those of a closely related strain.

bacteriophage: A virus that infects bacteria. Altered forms are used in DNA cloning work, where they are convenient vectors. Most of the bacteriophages are derived from two "wild" phages, called M13 and lambda (λ). Lambda phages are used to clone segments of DNA in the range of around 10-20 kb. They are lytic phages, i.e., they replicate by lysing their host cell and releasing more phages. On a bacteriological plate, this results in a small clear zone—a plaque. Some lambda vectors have also been developed, which are expression vectors. The M13 system can grow inside a bacterium, so that it does not destroy the cell it infects, but causes it to make new phages continuously. It is a single-stranded DNA phage, and is used for the Sanger di-deoxy DNA sequencing method. Both of these phages grow on *Escherichia coli* as a host bacterium.

bacteriostat: It is a biological or chemical agent that causes bacteriostasis. It stops bacteria from reproducing, while not necessarily harming them otherwise. Upon removal of the bacteriostat, the bacteria usually start to grow again. Bacteriostats are often used in plastics to prevent the growth of bacteria on the plastic surface. This is in contrast to bacteriocides, which kill bacteria.

Bacterium: It is a common name for the class Schizomycetes: minute (0.5-5 ìm), unicellular organisms, without a distinct nucleus. Bacteria are prokaryotes, and most of them are identified by means of Gram staining (q.v.). They are classified on the basis of their oxygen requirement as aerobic and anaerobic bacteria. Bacteria usually reproduce asexually by simple cell division, although a few undergo a form of sexual reproduction, termed *conjugation.* A few bacteria can photosynthesize (including green-blue cyanobacteria), some are saprophytes and others are parasites and can cause diseases. They are major agents of fermentation, putrefaction and decay, and frequently a source of contamination in tissue culture.

baculovirus: It is a class of insect virus which have been used to make DNA cloning vectors for gene expression in eukaryotic cells. Baculoviruses have a gene which is expressed at very high levels late in their infection cycle, filling the nucleus of the cell with many-sided bodies full of a protein, which is not needed to produce more viruses, but is necessary for the virus's spread in the wild.

baculovirus: It is an insect virus which forms the basis of a protein expression system.

baldwin effect: It refers to the hypothesized passing on of something learned, but not through the discredited Lamarckian theory of evolution (the inheritance of what previous generations experienced). Walking upright could be an example, as a band of our ancestors imitated some forgotten hominid who preferred that style of locomotion and then gave rise to descendants whose genes favored the behavior. The Baldwin Effect fills in a gap in how natural selection is thought to work by explaining how learnings normally invisible to it become innate.

bar diagram: It refers to the one-dimensional diagram where the length of the bar is important, and not the width. These are mainly of four types: (1) Simple, (2) Divided, (3) Percentage and (4) Multiple bar diagram.

base (nitrogenous): A cyclic, nitrogen-containing compound that is one of the essential components of nucleic acids. Exists in five main forms (adenine, A; guanine, G; thymine, T; cytosine, C; uracil, U). A and G have a similar structure and are called purines; T, C and U have a similar structure and are called pyrimidines. A base joined to a ribose sugar joined to a phosphate group is a nucleotide—the building block of nucleic acids.

base analogues: Unnatural purine or pyrimidine bases that differ slightly in structure from the normal bases, but can be incorporated into nucleic acids. They are often mutagenic.

base collection: As per Plant Genetic Resources, it is known as a collection of seed stock or vegetative propagating material (ranging from tissue cultures to whole plants) held for long-term security in order to preserve the genetic variation for scientific purposes and as a basis for plant breeding as multiplication and evaluation.

base pair (bp): The two strands that constitute DNA are held together by specific hydrogen bonding between purines and pyrimidines (A pairs with T; and G pairs with C). The size of a nucleic acid molecule is often described in terms of number of base pairs (symbol: bp) or thousand base pairs (kilobase pairs; symbol: kb; a more convenient unit) it contains.

base pair rule: Each type of base on one strand forms a bond with just one type of base on the other strand. This is called complementary base pairing. Here, purines form hydrogen bonds to pyrimidines, with A bonding only to T, and C bonding only to G. This arrangement of two nucleotides binding together across the double helix is called a base pair. In a double helix, the two strands are also held together via forces generated by the hydrophobic effect and pi stacking, which are not influenced by the sequence of the DNA. The two types of base pairs form different numbers of hydrogen bonds; AT forming two hydrogen bonds, and GC forming three hydrogen bonds—A rule followed by the nucleotides isolated from living cells. Erwin Chargaff, during his studies on nucleic acids, discovered that no matter what the biological source of the DNA, the ratio of the nucleotides obeyed this simple formula: % Adenine = % Thymine; % Guanine = % Cytosine This is called Chargaff's law, pieces of data of which allowed James Watson and Francis Crick to come up with the double-helix structure of DNA. The ratio arises because in normal cells, nearly all of the DNA is present as a complimentary double-helix, meaning that for each Adenine there is a corresponding Thymine, and for each Guanine, there is a corresponding Cytosine, and vice versa.

base pair: A pair of nitrogenous bases (a purine and a pyrimidine), held together by hydrogen bonds, that form the core of DNA and RNA i.e., the A:T, G:C and A:U interactions.

base substitution: Replacement of one base by another in a DNA molecule.

basophil: A type of white blood cell (leucocyte) produced by stem cells in the red bone marrow.

batch culture: A suspension culture in which cells grow in a finite volume of liquid nutrient medium and follow a sigmoid pattern of growth.

batch fermentation: A process in which cells or microorganisms are grown for a limited time. At the beginning of the fermentation, an inoculum is introduced into fresh medium, with no addition or removal of medium for the duration of the process.

bathymetric: It refers to scientific-sounding term for measuring something in deep water.

B-DNA: The normal form of DNA found in biological systems. It exists as a right-handed helix. There are ten base pairs per one full turn. The angle of rotation is 36°. Vertical rise is 3.4 A°.

bees algorithm: It is a population-based search algorithm first developed in 2005. It mimics the food foraging behaviour of swarms of honey bees. In its basic version, the algorithm performs a kind of neighbourhood search combined with random search and can be used for both combinatorial optimization and functional optimisation.

berm: It refers to a level, narrow ledge or bench that divides a stream or separates water from land, as in a canal path.

Bernoulli process: It is one where an experiment can result in only one of the two mutually exclusive outcomes such as success or failure, dead or alive, male or female, etc.

beta DNA (B form DNA): The normal form of DNA found in organisms, which exists as a right-handed helix.

Beta galactosidase: It is also called beta-gal or β-gal, is a hydrolase enzyme that catalyzes the hydrolysis of β-galactosides into monosaccharides.

beta lactamase: An enzyme that catalyses the formation of glucose and galactose from lactose.

beta lactamase: These are enzymes produced by some bacteria and are responsible for their resistance to beta-lactam antibiotics like penicillins, cephalosporins (are relatively resistant to beta-lactamase), cephamycins, and carbapenems (ertapenem). These antibiotics have a common element in their molecular structure: a four-atom ring known as a beta-lactam. The lactamase enzyme breaks that ring open, deactivating the molecule's antibacterial properties.

beta sheet: It refers to the three-dimensional arrangement taken up by polypeptide chains that consist of alternating strands linked by hydrogen bonds. The alternating strands together form a sheet that is frequently twisted. One of the secondary structural elements characteristic of proteins.

between sample variance: It is an estimate of the population variance derived from the variance among the sample means.

biased errors: They are those which arise because of bias in selection, estimation, etc.

bifunctional vector (shuttle vector): A deoxyribonucleic acid vector that is able to replicate in two different organisms, and therefore shuttles foreign nucleic acids between two different hosts.

bilateral symmetry: Body divisibility into mirror-image halves (right and left arms and legs, for example). Animals with bilateral symmetry display dorsal (top), ventral (bottom), anterior (front), and posterior (rear) orientations, whereas radial animals like starfish have only the dorsal and ventral. (Arms, legs, and wings evolved from the fins of lungfish swimming around over 400 million years ago.)

bimodal distribution: It is a distribution in which two values occur more frequently than the rest of the values in the data set.

binary vector system: A two-plasmid system in *Agrobacterium tumefaciens* for transferring into plant cells a segment of T-DNA that carries cloned genes. One plasmid contains the virulence gene (responsible for the transfer of T-DNA), and another plasmid contains the T-DNA borders, the selectable marker and the DNA to be transferred.

binomial distribution: It is a type of discrete distribution which describes the results of an experiment known as Bernoulli process.

binomial expansion: The probability that an event will occur 0, 1, 2, …, n times out of n is given by the successive terms of the expression $(p + q)n$, where p is the probability of the event occurring, and $q=1- p$.

binomial nomenclature: It is a system of classification developed by Carolus Linnaeus. According to it, in biology, each species is generally identified by two terms: the first is the genus to which it belongs, and the second is the specific epithet that distinguishes it from others in that genus (e.g., *Pisum sativum*, sweet pea). The genus name always has an initial capital; the specific epithet is never capitalized, even though it may be derived from a proper name. Both terms in the binomial are italicized.

bioaccumulation: In an organism, concentration of materials which are not critical components for that organism's survival. Usually, it refers to the accumulation of metals or other compounds (e.g., mercury, lead, arsenic, etc.). Many organisms—plants, fungi, protists, bacteria, etc.—accumulate metals when grown in a solution of them, either as part of their defence mechanism against the poisonous effect of those compounds, or as a side-effect of the chemistry of their cell walls.

bio-assay: A procedure for the assessment of a substance by measuring its effect in living cells or on organisms. Animals have been used extensively in drug research in bio-assays for the pharmacological activity of drugs. However, bio-assays are now usually developed using bacteria or animal or plant cells, as these are usually much easier to handle than whole animals or plants, are cheaper to make and keep, and avoid the ethical problems associated with the testing of animals. Sometimes used to detect minute amounts of substances that influence or are essential to growth.

bioaugmentation: Increasing the activity of bacteria that decompose pollutants; a technique used in bioremediation.

biocatalysis: Use of enzymes to catalyse chemical reactions.

biocenose (biocenosis): It refers to the interacting organisms living together in a habitat. A biotic community.

biocentric: It refers to putting the natural world, rather than the human world, into the perceived center of the cosmos. The land is not made for us: we are a part of it.

biochemical (biological) oxygen demand (BOD): It refers to the amount of oxygen required to dissolve and decompose organic matter. A water quality measurement often applied to treated sewage.

bioconcentration (biomagnification): It refers to the strengthening of a harmful and usually toxic substance as it moves up the food chain, as with DDT growing 400 times deadlier in seagulls and other carnivores than when first ingested by marsh animals.

biocontrol: It refers to using natural means like predators to control pests, like growing ginger to repel snails and slugs and nasturtiums to ward off aphids, which are also food for ladybugs and lacewing moths. Goldfish placed in water storage containers eat incoming mosquitos.

bioconversion: Conversion of one chemical into another by living organisms, as opposed to their conversion by enzymes (which is biotransformation) or by chemical processes. The usefulness of bioconversion is much the same as that of biotransformation, in particular its extreme specificity and ability to work in moderate conditions. However, bioconversion has several other properties, including the possibility of having several chemical steps. A major commercial application is in the manufacture of steroids. The "basic" steroid molecule, often isolated from plants, is itself a very complicated molecule, and not one that is easy to modify by normal chemical means to produce very specific molecules needed for drug use. However, a particular type of bioconversion that attacks only specific bits of the molecule can be used. Bioconversion is particularly useful for introducing chemical changes at specific points in large, complex molecules.

bioculture: It is an emerging area including by the medical sciences, social sciences, area studies, culture studies, biotechnology, disability studies, the humanities, and the economic and global environment. Bioculture defines the activity and consolidation of ideas created when the human intersects with the technological. Along these lines, one can see the biosphere—the earth as it is affected by the human—as the adaptation of the natural to the human and biocultures as the inter-adaptation of the human to the new technologies and ways of knowing characterized by the 21st century's attitude toward the body.

biodegradation: The breakdown of a compound by living organisms to its chemical constituents. Materials that can be easily biodegraded are colloquially termed *biodegradable.*

biodiversity: The variety of species (species diversity) or other taxa of animals, microorganisms and plants in a natural community or habitat, or of communities in a particular environment (ecological diversity), or of genetic variation in a species (genetic diversity). The maintenance of a high level of biodiversity is important for the stability of ecosystems.

biodynamic farming: It refers to alternative farming practice and philosophy introduced in eight lectures in Austria by Rudolf Steiner in 1924. Three of its techniques are: Horn manure preparation to introduce stirred nutrients into the soil, horn silica preparation to nourish plant growth, and compost preparations. Cycles such as those of the moon are also taken into account in order to build up humus and productivity. Alan Chadwick combined biodynamic methods with French intensive procedures in the late 1960s and founded an experimental farm and garden at UC Santa Cruz. (Word has it that he was a gifted horticulturalist but highly competitive.)

bioeconomics: It is the study of the dynamics of living resources using economic models. It is an attempt to apply the methods of environmental economics and ecological economics to empirical biology. Bioeconomics applies optimal control methods to mathematical models using environmental and ecological elements for resource protection issues relating to resource economics.

bioenergetics: The study of the flow and transformation of energy that occur in living organisms.

bioengineering: The use of artificial tissues, organs and organ components to replace parts of the body that are damaged, lost or malfunctioning.

bioenrichment: Adding nutrients or oxygen to increase microbial breakdown of pollutants.

bioethics: The branch of ethics that deals with the life sciences and their potential impact on society. At one extreme, it can be confronted; at the other extreme, it can become a name-calling argument between the "pro-biotechnology" and "antibiotechnology" schools of thought, which, as it reduces discussion to epithets and clichés, can make better sound bites.

biofuel: A gaseous, liquid or solid fuel that contains energy derived from a biological source. For example, rapeseed oil or fish liver oil can be used in place of diesel fuel in modified engines. A commercial application is the use of modified rapeseed oil, which as rapeseed methyl ester can be used in modified diesel engines, and is sometimes named biodiesel.

biogas (gobar gas): A mixture of methane and carbon dioxide resulting from the anaerobic decomposition of waste such as domestic, industrial and agricultural sewage.

biogenesis: The principle that a living organism can only arise from other living organisms similar to itself and can never originate from non-living material.

biohydrometallurgy: It is a method of involving microorganisms in the extraction of metals.

bioinformatics: It is an interdisciplinary branch dealing with the use and organization of information of biological interest. In particular, it is concerned with organizing bio-molecular databases; getting useful information out of such databases; utilizing powerful computers for analyzing such information; and integrating information from disparate biological sources.

bioinformatics: It refers to the field of endeavor that relates to the collection, organization and analysis of large amounts of biological data using networks of computers and databases (usually with reference to the genome project and DNA sequence information).

bioleaching (bio-oxidation): It is a process of extraction of metals from ores or soil by biological processes, mostly by microorganisms. It generally occurs by the dissolution of metals from their mineral source by naturally occurring microorganisms.

biolistics: It is a technique of inserting DNA molecules into cells. The DNA is mixed with small metal particles made up of tungsten or gold. These are then fired into a cell at a very high speed. They puncture the cell and carry the DNA into the cell. Biolistics has an advantage over transfection, transduction, etc., because it can apply to any cell, or indeed to parts of a cell.

biological containment: Restricting the movement of (genetically engineered) organisms by arranging barriers to prevent them from growing outside the laboratory.

biological control (biocontrol): It is the deliberate use of one organism to regulate the population size of a pest organism. There are three main branches of biological control.

biomagnification: It is also known as bioamplification or biological magnification. It is the increase in concentration of a substance that occurs in a food chain.

biomass: The total mass of cells or organisms produced by a population of living organisms. It is also referred to as organic mass that can be used either as a source of energy or for its chemical components.

biomechatronics: It is an applied interdisciplinary science that aims to integrate mechanical elements, electronics and parts of biological organisms. Biomechatronics includes the aspects of biology, mechanics, and electronics. It also encompasses the fields of robotics and neuroscience. One example of biomechatronics is a study done by Hugh Herr, a professor at M.I.T. Herr excised the muscles of frog legs to attach to a mechanical fish, and by pulsing electrical current through the muscle fibres, he caused the fish to swim. The goal of these experiments is to make devices that interact with human muscles, skeleton, and nervous system. The end result is that the devices will help human motor control that was lost or impaired by trauma, disease or birth defects.

biometry: It refers to the application of the statistical method or mathematical logic to the analysis and interpretation of biological variation.

biometry: The application of mathematical and statistical methods for the analysis of problems related to biology.

biomimicry (biomimetics): It is the examination of nature, its models, systems, processes, and elements to emulate or take inspiration from in order to solve human problems. The term *biomimicry* and *biomimetics* comes from the Greek words *bios*, meaning life, and *mimesis*, meaning to imitate. Other terms often used are bionics, bio-inspiration, and biognosis.

biopesticide: A compound that kills organisms by virtue of specific biological effect rather than as a broader chemical poison. *Bacillus thuringiensis* (Bt) toxin, which specifically interferes with the absorption of food from the guts of some insects, but is harmless to mammals, is an example.

biopharmaceuticals: These are the medical drugs produced using biotechnology. They are proteins, nucleic acids used for therapeutic or in vivo diagnostic purposes, and are produced by means other than direct extraction from a native biological source.

biopolymer: A large polymeric molecule produced by living organisms, e.g., protein, nucleic acid, polysaccharide, and lipid.

bioprocess: Any process that uses complete living cells or their components to effect the desired physical or chemical changes.

bioreactor: A tank or vessel in which cells, cell extracts or enzymes carry out a biological reaction primarily to produce either secondary metabolites or cells or somatic embryoids on a commercial scale, e.g., fermenter, fermentation vessel. Similarly, in case of filter bioreactor (mesh bioreactor), cells are grown on an open mesh of an inert material, which allows the culture medium to flow past it but retains the cells. This is similar in idea to membrane and hollow fibre reactors, but can be much easier to set up, being similar to conventional tower bioreactors, but with the meshwork replacing the central reactor space.

bioremediation: A process that uses living organisms to remove contaminants, pollutants or unwanted substances from soil or water.

biorobotics: It refers to the fields of cybernetics, bionics and even genetic engineering as a collective study. It is often used to refer to a real subfield of robotics: studying how to make robots that emulate or simulate living biological organisms mechanically or even chemically. The term is also used in a reverse definition: making biological organisms as manipulatable and functional as robots, or making biological organisms as components of robots.

biosensor: A device that uses an enzyme, antibiotic, organelle or whole cell as immobilized agent to detect or measure a chemical compound. A reaction between the immobilized agent and the molecule being analysed is transduced into an electric signal.

biotechnology: It is an interdisciplinary branch of science involving the use of biological processes or organisms for the production of materials and services of benefit to humankind. Biotechnology includes the use of techniques for the improvement of characteristics of economically important plants and animals and for the development of microorganisms to act on the environment. It also deals with the manipulation of living organisms, especially at the molecular genetic level, to produce new products, such as hormones, vaccines, monoclonal antibodies, interferons, interleukins, therapeutic proteins, etc.

biotic potential: It refers to a population's maximum production rate given ideal surroundings and resources.

biotic stress: Stress resulting from living organisms which can harm plants, such as viruses, fungi, bacteria, parasitic weeds and harmful insects. *cf* abiotic stress.

biotin: A vitamin of the B complex. It is a coenzyme for various enzymes that catalyse the incorporation of carbon dioxide into various compounds. It is essential for the metabolism of fats. Biotin is attached to pyruvate carboxylase by a long, flexible chain like that of lipoamide in the pyruvate dehydrogenase complex. Adequate amounts are normally produced by the intestinal bacteria in animals, a.k.a. vitamin H (in USA). The incorporation of a biotin-containing nucleotide into a DNA molecule is called biotin labelling.

biotinylated-DNA: A DNA molecule labelled with biotin by incorporation of biotinylated-dUTP into a DNA molecule. It is used as a non-radioactive probe in hybridization experiments, such as Southern transfer. The detection of labelled DNA is achieved by complexing it with streptavidin (an antibiotic with a high affinity for biotin) to which is attached a colour-generating agent such as horseradish peroxidase that gives a fluorescent green colour upon reaction with various organic reagents.

biotope: It refers to an environmentally uniform area. The physical aspect of an ecosystem.

biotransformation: The conversion of one chemical or material into another using a biological catalyst, usually an enzyme, or a whole, dead microorganism that contains an enzyme or several enzymes is called biotransformation. The term is nowadays used to describe the process of transformation of organisms by the introduction of foreign DNA.

bioturbation: It refers to when organisms disturb sediments, as with worms on the ocean floor that eat food stuck between the sand particles.

bioventing: It is an in situ remediation technology that uses indigenous microorganisms to biodegrade organic constituents adsorbed to soils in the unsaturated zone. The activity of the indigenous bacteria is enhanced by inducing air flow into the unsaturated zone.

bivalent: It is a pair of synapsed or associated homologous chromosomes during the process of genetic recombination in meiosis. Each duplicated chromosome comprises two chromatids. Thus, a bivalent comprises four chromatids.

bivariate analysis: If there is some relationship existing between the two variables, the statistical analysis of such data is called bivariate analysis.

black body: In theory, it refers to a body that absorbs and emits 100% of the electromagnetic radiation that strikes it and therefore appears black. Graphite comes close, with all but 3% absorption.

blot: It means the transfer of biomolecules like DNA, RNA or protein to an immobilizing matrix.

blunt (flush) end: The end of a DNA duplex molecule in which neither strand extends beyond the other.

blunt-end (ligation): It refers to the joining of DNA fragments that contain no overhang at either end and consequently no DNA bases available for hybridization (cf. sticky-end ligation).

blunt-end cut: To cleave phosphodiester bonds in the backbone of duplex DNA between the corresponding nucleotide pairs on opposite strands. This cleavage process results in both strands finishing at the same residue, i.e., there are no nucleotide extensions on either strand. a.k.a. flush-end cut.

blunt-end ligation: Ligation means joining of the nucleotides that are at the ends of two blunt-ended DNA duplex molecules.

bog (heath, muskeg): It refers to a marshy land covered by shrugs and mosses. Their acidic soils accumulate peat, the thick, carbonized vegetable tissue decomposed in water. The words "bogeyman" and "heathen" derive from outcasts who inhabited these poorly drained areas.

boid: It is a simulated individual particle or object moving about in a virtual space. In swarm intelligence the PSO algorithm uses a set of boids moving around in a search space to find an optimal solution for a given fitness function. In artificial life, boids typically refer to a collection of animated objects moving around in a virtual environment.

Boolean (logical) data type: It is the most primitive data type, having one of the two values (true or false) intended to represent the truth values of logic and Boolean algebra.

boring platform: It is a sterile bottom half of a Petri dish used for preparing explants with a cork borer.

bound water: Water held by the cell and not released if freezing occurs in the intercellular space.

bovine somatotrophin (BST): It is also known as bovine somatotropin a.k.a bovine growth hormone. This protein is found naturally in cattle. It is the earliest biopharmaceutical product using recombinant DNA technology.

bovine spongiform encephalopathy (BSE): It is commonly called mad cow disease and is caused by prions.

bridge: In plant tissue culture, when a liquid medium is used, filter paper or other substrate used as a wick and support structure to keep the explants firmly on the medium is called the bridge.

browning: In plant tissue culture, discoloration sometimes occurs due to phenolic oxidation of freshly cut surfaces of explant tissue. In later stages of culture, such discoloration may indicate a nutritional or pathogenic problem. It often leads to necrosis.

bubble column fermenter: A fermentation vessel, or bioreactor, in which the cells or microorganisms are kept suspended in a tall cylinder by rising air, which is introduced at the base of the vessel.

buffer: A solution that resists change in pH when an acid or alkali is added, or when the solutions are diluted.

buoyant density: The intrinsic density which a molecule, virus or sub-cellular particle has when suspended in an aqueous solution of salt, such as CsCl, or a sugar, such as sucrose. DNA from different species has a characteristic buoyant density, which reflects the proportion of G=C base pairs.

bush fallow: It refers to a secondary tropical or subtropical vegetation planted in forests in order to regenerate the soil between crops. In places like western Sudan, this traditional practice has been subverted by a Western emphasis on cash-producing monocrops in disregard of cultural or ecological consequences.

Buys-Ballot's Law: It refers to standing somewhere in the northern hemisphere with your back to the wind locates the low pressure area driving it on your left. The reverse is true in the southern hemisphere.

bycatch: It refers to accidental harvest of one organism instead of another, like the crustaceans caught in shrimp trawls and the dolphins trapped instead of tuna.

C

c (complementary) DNA: Synthesis of cDNAs is often the first step in cloning DNA sequences of interest. These are used as specific and sensitive probes in hybridization studies, because cDNAs usually do not include regulatory or other controlling sequences, and so they can be used to identify and isolate genes and their associated sequences from genomic DNA. Preparation of cDNA is done by the synthesis of double-stranded DNA complement of an mRNA sequence in vitro from a mature RNA template using reverse transcriptase (to create a single strand of DNA from the RNA template) and DNA polymerase (to create the double-stranded DNA).

CAAT box (CAT box): It is a distinct pattern of nucleotides with GGNCAATCT consensus sequence that occur upstream by 75-80 bases to the initial transcription site. The CAAT box signals the binding site for the RNA transcription factor, and is typically accompanied by a conserved consensus sequence. It is an invariant DNA sequence at about minus 70 base pairs from the origin of transcription in many eukaryotic promoters. Genes that have this element seem to require it for the gene to be transcribed in sufficient quantities. It is frequently absent from genes that encode proteins used in virtually all cells. This box along with the GC box is known for binding general transcription factors. CAAT and GC are primarily located in the region from 100-150bp upstream from the TATA box. Protein specific binding is required for the CCAAT box activation. These proteins are known as CCAAT box binding proteins/

CCAAT box binding factors. A CCAAT box is a feature frequently found before eukaryote coding regions, but is not found in prokaryotes.

caliche: It refers to the surface soil particles cemented together by lime (calcium carbonate). Lumps of it can block water, curtail root growth, and cause iron deficiency in the nearby plants.

callus (tissue culture): It is an unorganized mass of tissue formed during tissue culture on the artificial nutrient medium. Callus is generally actively dividing and undifferentiated. Callus cultures can be indefinitely maintained through regular subculturing. It is often induced by hormone treatment. It is also developed from injury (wounding) in the plants.

calorie (cal): It is equivalent to the amount of heat required to raise the temperature of 1 gm of water from 14.5°C to 15.5°C.

calorie: It refers to a quantity of energy equal to the amount of heat required to raise one gram of pure water from 14.5 to 15.5° Celsius under standard atmospheric pressure.

Cambrian explosion: It refers to supposed huge diversification of multicellular life forms in the earth's oceans during the Cambrian period 570 million years ago. All that's certain is that organisms living before the "explosion" did not leave behind many fossils.

cancer: It is a class of diseases in which a group of cells displays uncontrolled growth and intrusion on and destruction of adjacent tissues. It can sometimes spread to other locations in the body via lymph or blood. These three malignant properties of cancers differentiate them from benign tumours, which are self-limited, and do not invade or metastasize.

candidate gene: A gene whose function suggests that it may be involved in the genetic variation observed for a particular trait, e.g., the gene for growth hormone is a candidate gene for body weight. An experimental approach in which the knowledge of biochemistry of a trait is used to draw up a list of genes whose protein products could be involved in the trait is called candidate-gene strategy.

cap: The structure found on the 5?-end of eukaryotic mRNA, consisting of an inverted, methylated guanosine residue. The site in a gene where translation is initiated is called cap site.

capacitation: The final stage in the maturation process of a spermatozoon, taking place inside the female genital tract as the sperm penetrates the ovum.

capillary action: It refers to water movement through tiny absorbent channels, often against the force of gravity, made possible by water's firm hydrogen-oxygen bondings. Capillary action plays a major role in water diffusion through soils and organisms.

capsid: It is the protein shell of a virus. It consists of several oligomeric structural subunits made of protein called protomers. The observable 3-dimensional morphological subunits, which may or may not correspond to individual proteins, are called capsomeres. The capsid encloses the genetic material of the virus. Capsids are broadly classified according to their structure. The majority of viruses have capsids with either helical or icosahedral structure. Some viruses, such as bacteriophages, have developed more complicated structure.

carbohydrate: An organic compound performing many vital roles in living organisms. The simplest carbohydrates are sugars (saccharides), including glucose and sucrose. Polysaccharides are carbohydrates of much greater molecular weight and complexity; examples are starch, which serves as energy store in plant seeds and tubers; cellulose and lignin that form the cell walls and woody tissue of plants; glycogen, etc.

carbon flux: It refers to carbon movement; movement of organic compounds through an ecosystem. Specifically, the relationship between carbon dioxide absorbed by green plants and carbon dioxide respirated by various organisms.

carboxyl group: It refers to the COOH functional group, acidic in nature, found in all amino acids.

carboxypeptidases: Two enzymes (A and B) found in pancreatic juice. Their role is to remove the C-terminal aminoacid from a peptide; the A form removes any amino acid; the B form removes only lysine or arginine. Used when sequencing peptides.

carcinogen: A carcinogen may act on DNA causing dangerous changes, or it may work to increase the rate of cell division. This change in cell division may increase the probability of DNA changes. Some carcinogens promote the development of cancer in other ways as well. Some commonly known carcinogens include asbestos, radon, certain pesticides, arsenic, and tobacco smoke. Smokeless tobacco is a known carcinogen as well. One major carcinogen originates from something vital to life—the sun emits ultraviolet rays that are carcinogenic. Ultraviolet radiation is known to cause a variety of cancers affecting the skin. Hundreds of chemicals are capable of inducing cancer in humans or animals after prolonged or excessive exposure. There are many well-known examples of chemicals that can cause cancer in humans. The fumes of cadmium, nickel, and chromium are known to cause lung cancer. Vinyl chloride causes liver sarcomas. Exposure to arsenic increases the risk of skin and lung cancer. Leukemia can result from chemically induced changes in bone marrow from exposure to benzene and cyclophosphamide, among other toxicants. Other chemicals, including benzopyrene and ethylene dibromide, are considered by authoritative scientific organizations to be probably carcinogenic in humans because they are potent carcinogens in animals. Chemically-induced cancer generally develops many years after exposure to a toxic agent. A latency period of as much as thirty years has been observed between exposure to asbestos, for example, and incidence of lung cancer.

carcinoma: A malignant tumour derived from epithelial tissue, which forms the skin and the outer cell layers of internal organs.

carr: It refers to a wet area of deciduous scrub or woods grown from swampy soil.

carrier DNA: This is an additional DNA that increases the efficiency of transformation in electroporation and chemically-mediated DNA-delivery systems.

It is DNA of undefined sequence content, which is added to the transforming (plasmid) DNA used in physical DNA-transfer procedures.

carrier gas: The gas that carries the sample in gas chromatography.

carrier molecule: It is a lipid-soluble molecule that can bind to lipid-insoluble molecules and transport them across the membranes. Carrier molecules have specific sites that interact with the molecules they transport. Efficiency of carrier molecules may be modified by modifying the interacting sites through genetic engineering.

carrier: It is a term used in genetics to describe an individual that has one recessive mutant allele for some defective condition that is *masked* by a dominant normal allele at the same locus. For example, in case of recessive gene disorder like colour blindness, the person heterozygous to this disorder is only a carrier and will not suffer from the disease. However that person will be carrying the gene and is capable of transferring that gene to the next generation.

casein hydrolysate: Mixture of amino acids and peptides produced by enzymatic or acid hydrolysis of casein.

casein: A group of proteins found in milk.

casein: It refers to a tasteless white protein distilled from milk and used in dessert toppings, coffee whiteners, adhesives and binders, paint, and plastics. Sensitivity to it plays a role in milk allergies, Asperger's Syndrome, and autism.

catabolic pathway: A pathway by which an organic molecule is degraded in order to release energy for growth and other cellular processes, e.g., cellular respiration.

catabolite repression: Repression is the inactivation of certain sugar-metabolizing operons (e.g. *lac operon*) in favour of glucose utilization when glucose is the predominant carbon source in the cell.

catalysis: Catalysis is the process in which the rate of a chemical reaction is either increased or decreased by means of a chemical substance known as a catalyst. Unlike other reagents that participate in the chemical reaction, a catalyst is not consumed by the reaction itself. The catalyst may participate in multiple chemical transformations. Catalysts that speed the reaction are called positive catalysts. Catalysts that slow down the reaction are called negative catalysts or inhibitors. Substances that increase the activity of catalysts are called promoters, and substances that deactivate catalysts are called catalytic poisons.

catalyst: A substance that promotes a chemical reaction by lowering the activation energy of a chemical change.

catalytic RNA: An RNA molecule that brings about catalysis and cuts an RNA substrate.

cavitation: It refers to fast or even explosive erosion forced by air bubbles carried by a rapidly flowing liquid. Ruins its share of water pumps.

CD (cluster of differentiation) molecules: Group of antigens that is associated with a specific sub-population of T cells. There are designations for surface molecules on various cells of the immune system.

cDNA (complementary DNA): It refers to the DNA strand copied from mRNA using reverse transcriptase. A cDNA library represents all the expressed DNA in a cell.

cDNA clone: It is a double-stranded DNA molecule that is carried in a vector and was synthesized in vitro from an mRNA sequence by using reverse transcriptase and DNA polymerase.

cDNA cloning: A cDNA (short for complementary DNA or copy DNA) is a DNA copy of an RNA, usually an mRNA. Sometimes we want to make a cDNA library, a set of clones representing as many as possible of the mRNAs in a given cell type at a given time. Such libraries can contain tens of thousands of different clones. Other times, we want to make one particular cDNA—a clone containing a DNA copy of just one mRNA.

cDNA library: It is a collection of cloned cDNA (complementary DNA) fragments inserted into a collection of host cells, which together constitute some portion of the transcriptome of the organism.

cDNA library: It refers to the set of DNA fragments prepared from the total mRNA obtained from a selected cell, tissue.

(CDR) complementarity determining region: is a short amino acid sequence found in the variable domains of antigen receptor (e.g., immunoglobulin and T cell receptor) proteins that complement an antigen and therefore provide the receptor with its specificity for that particular antigen.

cell fusion: It is a process of bringing together two single cells usually of different species or genera to produce a hybrid cell in vitro. This is also called cell hybridization and is an important step in somatic hybridization.

cell line: A cell lineage that can be maintained in culture. A cell line arises from a primary culture. It implies that cultures from it consists of several lineages of the cells originally present in the primary culture.

cell suspension: Cells in culture in a liquid medium.

cellulosome: These are complexes of enzymes created by bacteria such as *Clostridium* and *Bacteroides*, but functioning outside the cell. They assist in digestion or degradation of molecules such as cellulose.

central dogma: The central dogma of molecular biology was first enunciated by Francis Crick in 1958. The central dogma of molecular biology deals with the detailed residue-by-residue transfer of sequential information. It states that information cannot be transferred back from protein to either protein or nucleic acid.

centrifugation: The procedure of separating molecules by size or density using centrifugal forces generated by a spinning rotor. G-forces of several hundred thousand times gravity are generated in ultracentrifugation. Centrifuge is a device in which solid or liquid particles of different densities are separated by rotating them in a tube in a horizontal circle. The denser particles tend to move along the length of the tube to a greater radius of rotation, displacing the lighter particles to the other end in a centrifuge.

chain terminator (terminator codon): Codons which do not code for an amino acid. They signal ribosomes to terminate protein synthesis. The codons are UAA(ochre), UAG (amber) and UGA (opal). Often two of these codons are found together at the end of a coding sequence of RNA.

chance selection: The selection of units depends entirely on chance, and one does not know beforehand which unit will actually constitute the sample.

chaparral: It refers to an evergreen shrub community adapted to dry seasons. Although it secrets a resin that burns like gasoline, homes continue to be built in its potentially explosive thickets.

chelation: It refers to chemical weathering in which chelates draw metallic cations (positively charged ions) out of the rocks and rocky minerals. Ultimately, all forms of weathering have a hand in forming soils.

chemical mutagen: A chemical or product capable of causing genetic mutation in living organisms exposed to it.

chemigation: It refers to dispensing a pesticide through an irrigation system. Sprinklers are often used for this.

chemiluminescence (chemoluminescence): It is the emission of light with limited emission of heat (luminescence), as the result of a chemical reaction. A standard example of chemiluminescence in the laboratory setting is found in the luminol test, where evidence of blood is taken when the sample glows upon contact with iron. When chemiluminescence takes place in living organisms, the phenomenon is called bioluminescence.

chemosterilant: It refers to a chemical that stops the pests from reproducing.

chemosynthesis: It refers to the chemical conversion of inorganic compounds found in an autotrophic organism's surroundings into food for it. (cf. photosynthesis.)

chemotaxis: Motion of a motile cell, organism or part towards or away from an increasing concentration of a particular chemical. It is the process by which two gametes often get attracted by each other especially.

chemotherapy: The treatment of disease, especially infections or cancer, by means of chemicals is called chemotherapy. Chemicals kill the malignant cells.

chernozem soils: Soils rich in humus and calcium, like the soils often seen in meadows and prairies.

chiasma: A visible point of junction between two non-sister chromatids of homologous chromosomes during the first meiotic prophase. It is at this point that the exchange of genetic material takes place to bring about genetic recombination.

chimera: It is the part of a plant with a genetically different constitution as compared with other parts of the same plant. It may result from different zygotes that grow together, or from artificial fusion. A recombinant DNA molecule that contains sequences from different organisms is called chimeric DNA, and a semi-synthetic gene, consisting of the coding sequence from one organism, fused to promoter and other sequences derived from a different gene is called a chimeric gene. Similarly, a fusion protein is called chimeric protein.

chimeric clone: It refers to the cloning artifact created by a foreign gene being inserted into a vector in an incorrect orientation resulting in the expression of a protein consisting of a fusion of two different gene products.

Chinampa: It refers to a maze of raised crop beds (camellones) on low, canal-fed islands of mud, clay, manure, and decomposing plants built up in lakes and ponds. Aztecs and Mayans used this farming method.

Chinook: It refers to a warm, dry wind that blows on the lee sides of mountains in North America.

chi-square distribution: It is a type of probability distribution, differentiated by their degrees of freedom, used to test a number of different hypotheses about variances, proportions and distributional goodness of fit.

chi-square test: A statistical test used to determine the significance of overall deviation between the observed and expected frequencies: $\chi^2 = \Sigma\ (fo - fe)^2 / fe$.

chi-squared test: A significance test used to statistically assess the goodness of fit of observed data to a prediction.

chitin: A nitrogenous polysaccharide occurring as skeletal material in many invertebrates and fungi.

chitin: It refers to an amino sugar and polysaccharide (an insoluble carbohydrate spun from interwoven simple sugars) found in some fungal cell walls and insect exoskeletons. Although abundantly produced—almost as much so as cellulose—some insecticides prevent it from cycling.

chlorofluorocarbons (CFCs): It refers to nonburning chemicals made of carbon, chlorine, and fluorine used in aerosol sprays, solvents, foams, refrigerants, and packing materials. When released into the air and exposed to ultraviolet radiation in the upper atmosphere, they form a gas that opens holes in the ozone layer.

chromat: It refers to the data file output from most popular DNA sequencers. Chromat files consist of the fluorescent traces generated by the sequencer for each of the four chemical bases, A, C, G, and T, together with the sequence and measures of the error in the traces at each sequence position.

chromatid: It contains the replicated DNA of each individual chromosome, which are joined by a centromere. It is actually one-half of two identical copies of a replicated chromosome. The two chromatids that are held together before separation during cell division are called sister chromatids.

chromatin: It consists of primarily DNA with some proteins like histones, and small amounts of RNA. Originally named because of the readiness with which it stains with certain dyes.

chromatography: It is the collective term for a set of laboratory techniques for the separation of mixtures. The mixture is dissolved in a fluid called the mobile phase, which carries it through a structure holding another material called the stationary phase. The various constituents of the mixture travel at different speeds, causing them to separate. The separation is based on differential partitioning between the mobile and stationary phases. Subtle differences in a compound's partition coefficient result in differential retention on the stationary phase and thus changing the separation. A variation is pseudo-affinity chromatography, in which a compound which is like a biological ligand is immobilized on a solid material, and enzymes or other proteins are bound to it.

chromosomal integration site: A chromosomal location where foreign DNA can be integrated, often without impairing any essential function in the host organism.

chromosomal mutation: A sudden heritable change in the gross structure of a chromosome, usually causing deleterious effects in the organism.

chromosomal polymorphism: The occurrence of one to several chromosomes in two or more alternative structural forms within a population.

chromosome banding: Staining of chromosomes in such a way that light and dark areas occur along the length of the chromosomes in repeatable patterns. Lateral comparisons identify pairs. Each chromosome can be identified by its banding pattern.

chromosome jumping: It is a popular technique in biotechnology that allows two segments of duplex DNA that are separated by thousands of base pairs (about 200 kb) to be cloned together. After sub-cloning, each segment can be used as a probe to identify cloned DNA sequences that, at the chromosome level, are roughly 200 kb apart.

chromosome walking: A technique that identifies overlapping cloned DNA fragments, which form one continuous segment of a chromosome. A series of colony hybridizations is then carried out, starting with some cloned fragment which has already been identified and is known to be in the region encompassed by the overlapping clones. This identified fragment is used as a probe to pick out clones containing adjacent sequences. These are then used as probes themselves to identify clones carrying sequences adjacent to them and so on.

chromosome: It is a term used to describe a single DNA molecule which is a tightly coiled strand of DNA and condensed into a compact structure by complexing with accessory histones. It is a structure usually seen when the cell is in the state of division.

circularization: It is a process associated with the replication of circular DNA. A DNA fragment generated by digestion with a single restriction endonuclease will have complementary 5' and 3' extensions (sticky ends). If these ends are annealed and ligated, the DNA fragment will convert to a covalently-closed circle.

cistron: The term was initially used by Benzer to describe a gene. It is being used even today, but not in the same sense as that of the original description (*cis-trans*). The term is used for a DNA sequence that codes for a specific polypeptide.

clinical trials: It refers to the research studies that involve patients. Biotechnology companies typically use clinical trials to assess the efficacy and safety of new therapies and to answer scientific questions. Typically, there are 3 phases during a clinical trial. Phase I is designed to evaluate the safety of the product in humans; phase II analyses the effects of dose escalation, and phase III definitively evaluates the clinical efficacy of the product.

cloche: It refers to a transparent plant cover that shields the plants from the cold.

clonal selection: The production of a population of plasma cells, all producing the same antibody in response to the interaction between a B lymphocyte producing a specific antibody and the antigen bound by that antibody.

clone: Clone literally means copy. A group of biomolecules or plasmids or cells or plants or animals that are identical are called clones. They are derived from single selected individual. Each individual is multiplied by a specific method of cloning, cloned gene or other sequence.

cloned line: A strain or line descended directly from a single individual clone.

cloning site: It is the insertion site on the chromosome or a plasmid at which a DNA sequence is incorporated.

cloning technique: To clone is to copy. Cloning is the technique of copying an organism or a gene. It is the process of producing similar populations of genetically identical individuals that occurs in nature when organisms such as bacteria, insects or plants reproduce asexually. Clones of adult animals are created by a process called somatic cell nuclear transfer. There are two variations of this method. They are the Roslin Technique and the Honolulu Technique. But in all of these techniques the resulting offspring will be genetically identical to the donor and not the surrogate, unless the donated nucleus is taken from a somatic cell of the surrogate. In this process, the nucleus of a somatic cell is removed and inserted into an unfertilized egg that has had its nucleus removed. The egg with its donated nucleus is then nurtured and divides until it becomes an embryo. The embryo is then placed inside a surrogate mother and develops inside the surrogate. After the first successful cloning of the fish, Asian carp in 1963 and the sheep Dolly in 1996 so for more than 20 animals have been cloned.

cloning vector (vehicle): A small DNA molecule—usually a plasmid or viral DNA chromosome—into which foreign DNA is inserted in the process of cloning genes or other DNA sequences of interest. It can carry inserted DNA and be perpetuated in a host cell.

closed continuous culture: It is one of the methods of suspension culture in which the inflow of fresh medium is balanced by outflow of corresponding volumes of spent medium. Cells are separated mechanically from the outflowing medium and added back to the culture.

cluster sampling: In such type of sampling, the population is divided into some recognizable subgroups, which are called clusters.

cluster: It refers to the grouping of similar objects in a multidimensional space. Clustering is used for constructing new features which are abstractions of the existing features of those objects. The quality of the clustering depends crucially on the distance metric in the space. In bioinformatics, clustering is performed on sequences, high-throughput expression and other experimental data. Clusters of partial or complete gene sequences can be used to identify the complete (contiguous) sequence and to better identify its function. Clustering expression data enables the researcher to discern patterns of co-regulation in groups of genes.

coconut milk (water): Liquid endosperm of the coconut, often used to supply organic nutrients to cultured cells and tissues. It significantly contains cytokinin, zeatin which was initially called CMF (coconut milk factor).

co-culture: A method in tissue culture in which two or more types of cells, such as a plant cell and a microorganism, or two types of plant cells are cultured.

coding regions (CDS): It refers to the portion of a genomic sequence bounded by start and stop codons that identifies the sequence of the protein being coded for by a particular gene.

coding sequence: It is the portion of a gene which directly specifies the amino acid sequence of its protein product.

coding strand: The strand of duplex DNA which contains the same base sequence found in the mRNA molecule resulting from transcription of that segment of DNA.

coding: The specification of a peptide sequence, by the code contained in DNA molecules, e.g., the codon AUG codes for the amino acid, methionine. *For details see the dictionary of genetic code in Appendix-2.*

codon optimization: An experimental strategy in which codons within a cloned gene—ones not generally used by the host cell translation system—are changed by in vitro mutagenesis to the preferred codons, without changing the amino acids of the synthesized protein.

codon: A set of three nucleotides in mRNA, functioning as a unit of genetic coding by specifying a particular amino acid during the synthesis of polypeptides in a cell. A codon specifies a transfer RNA carrying a specific amino acid, which is incorporated into a polypeptide chain during protein synthesis. The specificity for translating genetic information from DNA into mRNA, and then to protein, is provided by codon-anticodon pairing.

coefficient of correlation: The square root of the coefficient of determination. Its sign indicates the direction of the relationship between two variables, direct or inverse.

coefficient of variation: A relative measure of dispersion which expresses the standard deviation as a percentage of mean.

coenzyme: An organic molecule of low molecular weight and usually non-protein, such as a vitamin, that binds to an enzyme and promotes its catalytic activity.

co-evolution: It is the evolution of complementary adaptations in two species caused by the selection pressure that each exerts on the other. It is common in symbiotic associations, insect pollinated plants, etc.

cofactor: An organic molecule or inorganic ion necessary for the normal catalytic activity of an enzyme.

co-fermentation: The simultaneous growth of two microorganisms in one bioreaction.

cohesion: A force of attraction between like molecules leading to holding of a solid to a solid or a solid to a liquid.

cohesive ends: Double-stranded DNA molecules with single-stranded ends, which are complementary to each other, enabling different molecules to join each other.

co-integrate vector system: It is a two-vector plasmid system in which transferring of cloned genes to plant cells occurs. The cloning vector has a T-DNA segment that contains cloned genes. After introduction into *Agrobacterium tumefaciens*, the cloning vector DNA undergoes homologous recombination with a resident disarmed Ti plasmid to form a single plasmid carrying the genetic information for transferring the genetically engineered T-DNA region to plant cells.

colchicine: It is an alkaloid which inhibits spindle formation in cells during mitosis, so that chromosomes cannot separate during anaphase, thus inducing multiple sets of chromosomes. It is obtained from the tubers of a monocot *Colchicum autumnale.*

coleoptiles: Protective sheath covering the shoot apex of the embryo in monocotyledenous plants. Coleoptiles of *Avena sativa* were used by Went in the discovery of auxins.

coleorhizae: A protective sheath surrounding the radicle of monocotyledenous plants.

co-linearity: A relationship in which the units in one molecule occur in the same sequence as the units in another molecule which they specify, e.g., the nucleotides in a gene are co-linear with the amino acids in the polypeptide encoded by that gene.

colony hybridization: A technique that uses a nucleic acid probe to identify a bacterial colony with a vector carrying a specific cloned gene or genes.

colony: An aggregate of identical cells. For example, a bacterial colony is defined as a visible cluster of bacteria growing on the surface of or within a solid medium, presumably cultured from a single cell.

combinatorial chemistry: It refers to the use of chemical methods to generate all possible combinations of chemicals starting with a subset of compounds. The building blocks may be peptides, nucleic acids or small molecules. The libraries of compounds formed by this methodology are used to probe for new pharmaceutical reagents (see high-throughput screening).

combinatorial library: During the ligation reaction with cDNAs of light and heavy antibody chains into a bacteriophage lambda vector, many novel combinations consisting of one heavy and one light chain coding region are formed. The combinatorial library comprises these combinations, each in a separate vector.

commensalism: It refers to a co-evolutionary relationship between the species (usually animal species) that benefits one without significantly impacting the other.

chlorophyll: It refers to a green, sunlight-capturing pigment in plants and some bacteria. (See Photosynthesis.)

commensalism: The interaction of two or more dissimilar organisms where the association is advantageous to one without affecting the others.

commodification (commoditization): It is the transformation of goods and services (or things that may not normally be regarded as goods or services) into a commodity.

comparative gene mapping: The comparison of map locations of genes between the species.

comparative positional candidate gene: A gene that is likely to be located in the same region as a DNA marker that has been shown to be linked to a single-locus trait or a quantitative trait locus.

competent cell: A competent cell is the one capable of developing into a fully functional embryo.

competitive exclusion: It refers to a phenomenon, where one species competes another into extinction.

complement proteins: Proteins that bind to antibody-antigen complexes.

complementarity (in DNA): Thymine in one strand always pairs with adenine in the other strand, and cytosine in one strand always pairs with guanine in the other strand. Thus, the two strands of a double helix DNA are held together, and it is called complementarity.

complementary determining region (CDR): It refers to the hypervariable regions of an antibody molecule, consisting of three loops from the heavy chain and three from the light chain, that together form the antigen-binding site.

complementary genes: Two or more interdependent genes, such that the dominant allele from either gene can only produce an effect on the phenotype of an organism if the dominant allele from the other gene is also present. This phenomenon is called complementary gene interaction.

complementary homopolymeric tailing: The process of adding complementary nucleotide extensions to different DNA molecules.

complementary nucleotides: Members of the pairs adenine-thymine, adenine-uracil, and guanine-cytosine.

complete digest: Complete cleaving of the potential target sites within that DNA preparation by an endonuclease.

complexity (of gene sequence): This term may be thought of as synonymous with regions of locally biased amino acid composition. In these regions, the sequence composition deviates from the random model that underlies the calculation of the statistical significance (P-value) of an alignment. Such alignments among low complexity sequences are statistically but not biologically significant, i.e., one cannot infer homology (common ancestry) or functional similarity.

composite transposon: A transposable element formed when two identical or nearly identical transposons insert on either side of a non-transposable segment of DNA.

compound chromosome: A chromosome formed by the union of two separate chromosomes.

concatemer: It is a long continuous DNA molecule that contains multiple copies of the same DNA sequences linked in series. These polymeric molecules are usually copies of an entire genome linked end to end and separated by *cos* sites (a protein binding nucleotide sequence that occurs once in each copy of the genome). Concatemers are frequently the result of rolling circle replication, and may be seen in the late stage of bacterial infection by phages. As an example, if the genes in the phage DNA are arranged ABC, then in a concatemer the genes would be ABCABCABCABC.

concordance: In genetics, the presence of same trait in both the members of a pair of twins (or set of individuals).

conditional lethal mutation: A mutation that is lethal under one set of environmental conditions.

Conditional probability: The probability of one event occurring, given that another event has occurred.

conditioning: The effects on phenotypic characters of external agents during critical developmental stages.

configuration (in software): It refers to the complete ordering and description of all parts of a software or database system. Configuration management is the use of software to identify, inventory and maintain the component modules that together comprise one or more systems or products.

conformation: It refers to the precise three-dimensional arrangement of atoms and bonds in a molecule describing its geometry and hence its molecular function.

conjugation: The unidirectional transfer of DNA from one bacterium to another, involving cell-to-cell contact. The plasmid usually encodes the majority of functions necessary for its own transfer.

conjugative functions: Plasmid-based genes and their products that facilitate the transfer of a plasmid from one bacterium to another.

conjunctive symbiosis: It refers to a mutually beneficial relationship in which the two participants join into a single organ or body.

connectance: It refers to the actual food web interactions between the species compared with the total possible number; usually expressed as a fraction.

consanguinity: It refers to the property of being from the same lineage as another person. In that respect, consanguinity is the quality of being descended from the same ancestor as another person. Consanguinity is an important legal concept in that the laws of many jurisdictions consider consanguinity as a factor in deciding whether two individuals may be married, or whether a given person inherits the property when a deceased person has not left a will.

consensus sequence: It refers to the single sequence delineated from an alignment of multiple constituent sequences that represents a "best fit" for all those sequences. A "voting" or other selection procedure is used to determine which residue (nucleotide or amino acid) is placed at a given position in the event that not all of the constituent sequences have the identical residue at that position.

consensus sequence: The nucleotide sequence that is present in the majority of genetic signals or elements that perform a specific function.

constant domains: Regions of antibody chains that have the same amino acid sequence in different members of a particular class of antibody molecules.

constitutive enzyme: An enzyme that is synthesized continually regardless of growth conditions.

constitutive gene: A gene that encodes a product required in the maintenance of basic cellular processes or cell architecture. Also known as housekeeping gene.

constitutive promoter: An unregulated promoter that allows for continual transcription of its associated gene.

constitutive synthesis (expression): It refers to the synthesis of mRNA and protein at an unchanging or constant rate regardless of a cell's requirements (see housekeeping genes).

constitutive synthesis: Continual production of RNA or protein by an organism.

constitutive: An organism is said to be constitutive for the production of an enzyme or other protein if that protein is always produced by the cells under all physiological conditions.

contaminant: An unwanted organism (bacteria or fungus) accidentally introduced into a culture medium. Contaminant may overgrow the plant cells and consequently inhibit their growth.

contig: A contig is a set of overlapping DNA segments derived from a single genetic source. A contig in this sense can be used to deduce the original DNA sequence of the source. The alignment of sequence data from large, adjacent regions of the genome to produce a continuous nucleotide sequence across a chromosomal region is called contig mapping.

contig: It refers to the length of contiguous sequence assembled from partial, overlapping sequences, generated from a "shotgun" sequencing project. Contigs are typically created computationally, by comparing the overlapping ends of several sequencing reads generated by restriction enzyme digestion of a segment of genomic DNA. The creation of contigs in the presence of sequencing errors, ambiguities and the presence of repeats is one of the most computationally challenging aspects of the role of bioinformatics in genome analysis.

contingency table: A table having R rows and C columns. Each row corresponds to a level of one variable, each column to a level of another variable. Entries in the body of the table are the frequencies with which each variable combination occurred.

continuous culture: A suspension culture continuously supplied with nutrients by the inflow of fresh medium. The culture volume is normally constant. Turbidostat and chemostat are the two major techniques of continuous culture.

continuous data: Data that may progress from one class to the next without break and may be expressed by either whole numbers or fractions.

continuous fermentation: A process in which the cells or microorganisms are maintained in culture in the exponential growth balanced by the removal of cell suspension from the bioreactor.

continuous probability distribution: A probability distribution in which the variable is allowed to take any value within a given range.

continuous variable: Variables which can take any value within the certain range exhibited by the population.

control: A standard protocol or treatment for comparison with the experimental treatment.

controlled environment: The environment in which parameters, such as light, temperature, relative humidity and sometimes the partial gas pressure, are fully controlled.

convenience sampling: Selection of items results in obtaining a chunk of the population. Chunk is a convenient slice of a population which is commonly referred to as a sample.

convergence: It refers to the end-point of any algorithm that uses iteration or recursion to guide a series of data processing steps. An algorithm is usually said to have reached convergence when the difference between the computed and observed steps falls below a predefined threshold.

convergent evolution: It refers to the evolution of similar characteristics in widely separated populations. Wings in unrelated species of birds are an example.

coordinate repression: Correlated regulation of the structural genes in an operon.

coordinate: Together the ordinates and abscissa are called coordinate to the point.

copolymers: Mixtures consisting of more than one monomer.

copy number: The average number of molecules of a plasmid or gene.

corepressor: It is a protein that decreases gene expression by binding to a transcription factor which contains a DNA binding domain. The corepressor is unable to bind DNA by itself. The corepressor can repress transcriptional initiation by recruiting histone deacetylases, which catalyze the removal of acetyl groups from lysine residues. This increases the positive charge on histones, which strengthens in the interaction between the histones and DNA, making the latter less accessible to transcription.

correlation analysis: It is a technique to ascertain the strength of relationship between two variables.

correlation and regression: These techniques help in measuring the independence or relationship between bivariate data and predict the value of one variable for the given value of other variable.

correlation: It is a statistical tool which measures the closeness of the relationship between two variables.

cos ends: The 12-base, single-strand, complementary extensions of bacteriophage lambda DNA.

cos sites(ends): A 12-nucleotide base-long segment of single-stranded DNA that exists at both ends of the bacteriophage lambda's double-stranded genome. The two *cos* sites at the ends of the genome are complementary to one another, so that the genome can become circular once the virus has infected a host bacterium. The circular genome can then be duplicated continuously until there are many repeats of it strung together. The *cos* sites show the virus where to cut them apart right before they are packaged into individual capsids as new progeny viruses ready to infect more host cells.

co-segregation: The transmission of two or more linked genes on a chromosome to the same daughter cell leading to the inheritance by the offspring of these genes together.

cosmid: It is a type of hybrid plasmid that contains *cos sequences*, DNA sequences originally from the Lambda phage. Cosmids can be used to build genomic libraries. Cosmids are able to contain 37 to 52 kbp of DNA, while normal plasmids are able to carry only 1-20 kbp. They can replicate as plasmids if they have a suitable origin of replication, for example, SV40 ori in mammalian cells, ColE1 ori for double-stranded DNA replication, or fl ori for single-stranded DNA replication in prokaryotes. They frequently also contain a gene for selection such as antibiotic resistance, so that the transfected cells can be identified by plating on a medium containing the antibiotic. Those cells which did not take up the cosmid would be unable to grow. Unlike plasmids, they can also be packaged in phage capsids, which allows the foreign genes to be transferred into or between, cells by transduction. Plasmids become unstable after a certain amount of DNA has been inserted into them, because their increased size is more conducive to recombination. To circumvent this, phage transduction is used instead. This is made possible by the *cohesive ends*, also known as *cos* sites. In this way, they are similar to using the lambda phage as a vector, but only that *all* the lambda genes have been deleted with the exception of *cos* sequence.

cosmids: It refers to the DNA vectors that allow the insertion of long fragments of DNA (up to 50 kbases).

cot curve: Cot curve is concerned with the measurement of degree of reannealing of DNA strands. It is a curve drawn with X-axis having DNA concentration unit multiplied by time. Since the initial concentration is considered represented as Co, so when multiplied with time t, it becomes "Cot" and the graph is known as Cot curve. The graph is drawn against % re-annealed versus Cot.

co-transfection: Simultaneous transfection with two separate, unrelated nucleic acid molecules, one of which may contain a gene that is easily assayed and acts as a marker.

co-transformation: The simultaneous transformation of two bacterial marker genes.

coupling: The phase state in which either two dominant or two recessive alleles of two different genes occur on the same chromosome. The term was first used to describe linkage along with repulsion. Earlier coupling-repulsion was proposed in 1911 by Bateson and Punnett to explain linkage.

covalently closed circle (CCC): A circular molecule of double-stranded DNA which is supercoiled, or coiled up on itself due to internal tensions, because there are no breaks in the phosphate backbone (upon which the nucleotide bases are mounted) to relieve the tensions and allow it to form an open circle.

cpDNA: DNA of plant plastids.

crown gall: A bulbous growth that occurs at the base of certain plants as a result of infection by *Agrobacterium tumefaciens*.

cryogenic: At very low temperature.

cryopreservation: The preservation or storage of plant seeds and pollen, micro-organisms, animal sperm, and tissue culture cell lines. It is a process. Preservation is done by cooling to low sub-zero temperatures, such as (typically) 77 K or -196°C (the boiling point of liquid nitrogen). At these low temperatures, any biological activity, including the biochemical reactions that would lead to cell death, is effectively stopped. However, when vitrification solutions are not used, the cells being preserved are often damaged due to freezing during the approach to low temperatures or warming to room temperature.

cryoprotectant: Compound preventing cell damage during freezing and thawing processes. Two types of cryoprotectant agent, permeating and non-permeating, are commonly used.

cryosphere: It refers to the part of the earth that remains below the freezing point.

cryoturbation: It refers to frost churning of soils.

cryptophyte: It refers to a plant whose reproductive organs (e.g., bulbs, corms) are underground.

cultivar: It refers to an organism with desirable breeding qualities.

crystal structure: The term is used to describe the high resolution molecular structure derived by x-ray.

cultivar(cv): A cultivated variety of plant.

culture room: It is also called growth room and is used for maintaining cultures often in a controlled environment.

culture: A population of plant or animal cells or microorganisms grown under controlled conditions.

cumulative frequency distribution: A tabular display of data showing as to how many observations lie above or below certain values.

curing: It refers to various food preservation and flavouring processes, especially of meat or fish, by the addition of a combination of salt, sugar, nitrates or nitrite. It also means the elimination of a plasmid from its host cell. Many agents which interfere with DNA replication, e.g., ethidium bromide, can cure plasmids from either bacterial or eukaryotic cells.

curvilinear relationship: An association between two variables that is described by a curved line.

cyborg (cybernetic organism): It is a being with both biological and artificial parts.

cybrid: It is a cytoplasmic hybrid produced by the fusion of a protoplast with a whole cell. The nucleus and cytoplasm of the fused cell products are from different genetic sources.

cyclic AMP (cyclic adenosine monophosphate): A messenger that regulates many intracellular reactions by transducing signals from extracellular growth factors to cellular metabolic pathways.

cyclodextrin (cycloamyloses): They make up a family of cyclic oligosaccharides, composed of 5 or more α-D-glucopyranoside units linked 1->4, as in amylose (a fragment of starch). The 5-membered macrocycle is not natural. Recently, the largest well-characterized cyclodextrin contains 32 1,4-anhydroglucopy-ranoside units, while as a poorly characterized mixture, even at least 150-membered cyclic oligosaccharides are also known.

cyclone: It refers to a low-pressure centre wrapped in rotating movements of air.

cyst: It refers to a capsule-like sac that encloses a spore, parasite, or abnormal growth.

cylons (Cybernetic Lifeform Node): They are a race of sentient machines, created by humans, which appear in the reimagined *Battlestar Galactica* television series and its prequel spin-off *Caprica*. They have several different forms, some of which perfectly resemble and even mimic the behaviour of humans, while others are more mechanical in appearance and function.

cysteine: It is a semi-essential amino acid, i.e., it can be biosynthesized in humans.

cytokines: They are any of a number of substances secreted by specific cells of immune system, which carry signals locally between the cells, and thus have an effect on other cells. They are a category of signalling molecules that are used extensively in cellular communication, e.g., proteins, peptides, or glycoproteins. The term cytokine encompasses a large and diverse family of polypeptide regulators that are produced widely throughout the body by the cells of diverse embryological origin. Basically, the term "cytokine" has been used to refer to immunomodulating agents. Conflicting data exists about what is termed a cytokine and what is termed a hormone. Anatomic and structural distinctions between the cytokines and classic hormones are fading as we learn more about each. Classic protein hormones circulate in nanomolar concentrations that usually vary by less than one order of magnitude.

cytokinesis: Cytoplasmic division and other changes exclusive of nuclear division that are a part of mitosis or meiosis.

cytokinin: The compounds derived from adenine and used as plant growth regulators (hormones). These are characterized as substances that induce cell division and cell differentiation.

cytoplasm: The living material of the cell.

cytoplasmic genes (plasmogenes): Genes found in DNA-containing bodies in the cell but external to the nucleus.

cytoplasmic inheritance: Hereditary transmission depend on the cytoplasm or structures in the cytoplasm rather than the nuclear genes—extra-chromosomal inheritance. Thus, plastid characteristics in plants are inherited by a mechanism independent of nuclear genes.

cytoplasmic male sterility (CMS): It is the total or partial male sterility associated with plant biology as a result of specific nuclear and mitochondrial interactions. Male sterility is the failure of plants to produce functional anthers, pollen, or male gametes.

cytosol: The fluid portion of the cytoplasm.

cytotoxic T cell (T_C, CTL, T-Killer cell, cytolytic T cell, CD8+ T-cells or killer): They are the cells of immune system capable of inducing the death of infected somatic or tumour cells. They kill cells that are infected with viruses or otherwise damaged or dysfunctional.

D

dabs: These are the antibodies with only one protein chain derived from only one of the domains of the antibody structure.

dalton (Da): It is a unit of atomic mass roughly equivalent to the mass of a hydrogen atom. The unified atomic mass unit or atomic mass unit (u), or dalton or, sometimes, universal mass unit (u), is a unit of mass used to express atomic and molecular masses. It is the approximate mass of a hydrogen atom, a proton, or a neutron.

Darwinian cloning: It is a selection of a clone from a large number of essentially random starting points, rather than isolating a natural gene or making a carefully designed artificial one.

Darwinian fitness (*w*): It is a central idea in evolutionary theory. It can be defined either with respect to a genotype or a phenotype. In either case, it is equal to the average contribution to the gene pool of the next generation that is made by an average individual of the specified genotype or phenotype. If differences between the alleles at a given gene affect fitness, then the frequencies of the alleles will change over generations; the alleles with higher fitness become more common. This process is called natural selection.

data array: The arrangement of raw data by observations in either ascending or descending order.

data cleaning: It refers to the process whereby automated or semi-automated algorithms are used to process experimental data, including noise, experimental errors and other artifacts, in order to generate and store high-quality data for use in subsequent analysis. Data cleaning is typically required in high-throughput sequencing where compression or other experimental artifacts limit the amount of sequence data generated from each sequencing run or "read."

data mining: It refers to the ability to query very large databases in order to satisfy a hypothesis ("top-down" data mining), or to interrogate a database in order to generate new hypotheses based on rigorous statistical correlations ("bottom-up" data mining).

data point: A single observation from a data set.

data processing: It is defined as the systematic performance of operations upon data such as handling, merging, sorting, and computing. The semantic content of the original data should not be changed, but the semantic content of the processed data may be changed.

data set: A collection of data.

data warehouses: It refers to the vast arrays of heterogeneous (biological) data, stored within a single logical data repository, that are accessible to different querying and manipulation methods.

data: A collection of any number of related observations on one or more variables.

database: It refers to any file system by which data gets stored following a logical process.

deadhead: It refers to removal of spent blossoms in order to encourage new flowers.

deceleration phase: The phase of declining growth rate, following the linear phase and preceding the stationary phase in most batch-suspension cultures.

deconvolution: It refers to the mathematical procedure to separate out the overlapping effects of molecules such as mixtures of compounds in a high-throughput screen, or mixtures of cDNAs in a high density array.

de-differentiation: The process by which the cells lose their specialization and proliferate by cell division to form a mass of cells, especially during plant tissue culture under the influence of growth regulators. This process often leads to the formation of callus.

deficiency: Insufficiency or absence of one or more usable forms of enzymatic, nutritional or environmental requirements.

degeneracy of the genetic code: The specification of one amino acid by more than one codon. Almost all amino acids, except methionine and tryptophan, are coded by more than one codon.

dehalogenation: It is the process of removing chlorine molecules from a contaminant in the soil. Contaminated soil is screened, processed with a crusher, and mixed with chemicals (i.e., reagents). The mixture is heated in a reactor. Dehalogenation (i.e., dechlorination) is achieved either by replacing the chlorine molecules or by partially decomposing and partially volatizing contaminants.

dehydrogenation: A chemical reaction in which hydrogen is removed from a compound.

de-ionized water: Water which is free of most inorganic and organic compounds.

deletion mapping: It refers to the process in which different deletions in a region of DNA are created and used to map the functionally critical areas of that DNA. For example, the minimal region of DNA required for a test promoter can be ascertained by systematic deletions in the region of interest.

deletion: It is the removal of a small length of DNA sequence from the chromosomes or the mutation involving the removal of one or more base pairs in DNA sequence. Deletions can be caused by errors in chromosomal crossover during meiosis. They may cause several serious genetic diseases or they may not cause any harm at all. Deletions also cause frameshift.

deletion: It refers to the chromosomal alteration in which a portion of the chromosome or the underlying DNA is lost.

denaturated DNA: It is also called DNA melting. It is the process by which double-stranded deoxyribonucleic acid unwinds and separates into single-stranded strands through the breaking of hydrogen bonding between the bases. Both the terms are used to refer to the process as it occurs when a mixture is heated, although "denaturation" can also refer to the separation of DNA strands induced by chemicals like urea. For multiple copies of DNA molecules, the melting temperature (T_m) is defined as the temperature at which half of the DNA strands are in the double-helical state and half are in the "random-coil" state. The melting temperature depends on both the length of the molecule, and specific nucleotide sequence composition of that molecule.

denaturation: Loss of native configuration of a macro-molecule (protein or nucleic acid) by physical or chemical means, usually accompanied by loss of biological activity by the induction of structural alterations disrupting the biological activity of a molecule. Often breaking hydrogen bonds between base pairs in double-stranded nucleic acid molecules to produce single-stranded polynucleotides, or altering the secondary and tertiary structure of a protein, thus destroying its activity.

dendrogram: It refers to the graphical procedure for representing the output of a hierarchical clustering method. A dendrogram is strictly defined as a binary tree with a distinguished root, that has all the data items at its leaves. Conventionally, all the leaves are shown at the same level of drawing. The ordering of the leaves is arbitrary, as is their horizontal position. The heights of the internal nodes may be arbitrary, or may be related to the metric information used to form the clustering.

denitrification: A chemical process in which nitrates in the soil are reduced to molecular nitrogen.

denitrification: Chemical conversion of dissolved nitrogen (nitrite) into gaseous nitrogen. Fires on particular soils lead to dentrification.

density gradient centrifugation: High-speed centrifugation in which molecules float at a point where their density equals to that in a gradient of caesium chloride or sucrose.

deoxyribonuclease (DNase): The enzyme that hydrolyses DNA.

dependable variable: The variable which is to be predicted in regression analysis.

de-repression: The process of "turning on" the expression of a gene.

derivative: Resulting from or derived from.

desalinization plant: It refers to a site where seawater is boiled to separate steam from salt; the steam condenses into fresh water. Since this requires a large amount of energy, normally provided by burning coal or oil, desalinization is normally used only in very dry areas.

desiccant: The compound used to remove moisture or water.

desiccate: To exhaust or deprive of water or moisture by drying. Any chemical used for this purpose is called a desiccant. An apparatus used for drying and preventing hygroscopic samples from rehydrating is a desiccator. The process is desiccation.

designer baby: It refers to a baby whose genetic makeup has been artificially selected by genetic engineering combined with in vitro fertilisation to ensure the presence or absence of particular genes or characteristics. The term is derived by comparison with "designer clothing". It implies the ultimate commodification of children and is therefore usually used pejoratively to signal opposition to such use of reprogenetics.

dessicator: Apparatus for drying or depriving of moisture.

desulphurization: The method for removing sulphur from oil and coal by the use of bacteria.

detergent: The chemical often used in the manufacturing of washing or cleaning substances, which actually lowers the surface tension of a solution, thus improving its cleaning properties.

determinate growth: Limited growth in time.

detritus: It refers to decomposing organic matter.

dextrin: An intermediate polysaccharide compound resulting from the hydrolysis of starch to maltose by amylase enzymes.

diabetes: *Diabetes mellitus* is a disease associated with the absence or reduced levels of insulin, which is a hormone essential for the process of glycogenesis, in which glucose is converted into glycogen. *Diabetes insipidus* is another disease caused by the deficiency of anti-diuretic hormone (ADH).

diagnostic procedure: A test by which the presence or absence of an pathogen or a toxic substance or nucleic acid sequence alteration can be determined.

diakinesis: A stage of meiosis in which the separation of homologous chromosomes is almost completed and they start moving towards the opposite poles.

diazotroph: They are bacteria that fix atmospheric nitrogen gas into a more usable form such as ammonia. A diazotroph is an organism that is able to grow without external sources of fixed nitrogen. All diazotrophs contain iron-molybdenum nitrogenase systems. Two of the most studied systems are those of *Klebsiella pneumoniae* and *Azotobacter vinelandii*. These systems are used because of their genetic tractability and fast growth. Diazotrophs are scattered across bacterial taxonomic groups. Even within a species that can fix nitrogen, there may be strains that do not fix nitrogen. Fixation is shut off when other sources of nitrogen are available, and, for many species, when oxygen is at high partial pressure.

dicentric chromosome: Chromosome having two centromeres.

di-deoxynucleotide (ddNTPs): These are nucleotides lacking a 3'-hydroxyl (-OH) group on their deoxyribose sugar. Since deoxyribose already lacks a 2'-OH, dideoxyribose lacks hydroxyl groups at both its 2' and 3' carbons. The lack of this hydroxyl group means that, after being added by a DNA polymerase to a growing nucleotide chain, no further nucleotides can be added as no phosphodiester bond can be created based on the fact that deoxyribonucleoside triphosphates (which are the building blocks of DNA) allow DNA chain synthesis to occur through a condensation reaction between the 5' phosphate (following the cleavage of pyrophosphate) of the current nucleotide with the 3' hydroxyl group of the previous nucleotide. The dideoxyribonucleotides do not have a 3' hydroxyl group, hence no further chain elongation can occur once this dideoxynucleotide is on the chain. This can lead to the determination of DNA sequence. Thus, these molecules form the basis of dideoxy chain-termination method of DNA sequencing, which was developed by Frederick Sanger in 1977.

differential centrifugation: It is a common procedure used in microbiology and cytology to separate certain organelles from whole cells for further analysis of specific parts of cells. In the process, a tissue sample is first homogenised to break the cell membranes and mix-up the cell contents. The homogenate is then subjected to repeated centrifugations, each time removing the pellet and increasing the centrifugal force. Finally, purification may be done through equilibrium sedimentation, and the desired layer is extracted for further analysis. Separation is based on size and density, with larger and denser particles pelleting at lower centrifugal forces. As an example, unbroken whole cells will pellet at low speeds and short intervals such as 1000g for 5 minutes. Smaller cell fragments and organelles remain in the supernatant and require more force and greater times to pellet. In general, one can enrich for the following cell components, in the separating order in actual application as: Whole cells and nuclei; mitochondria, lysosomes and peroxisomes; microsomes (vesicles of disrupted endoplasmic reticulum); and ribosomes and cytosol.

differentially permeable: It refers to a membrane, through which different substances diffuse at different rates; some substances may be unable to diffuse through such a membrane.

diffusion: It is the movement of molecules from a region of higher concentration to a region of lower concentration.

digest: To cut or cleave DNA molecules with one or more restriction endonucleases.

dihaploid: It is an individual arising from a doubled haploid.

dihybrid: It is the progeny of a cross between homozygous parents differing at two loci; an individual that is heterozygous for two pairs of alleles.

dimer: A dimer is a chemical or biological entity consisting of two structurally similar subunits called monomers, which are joined by bonds that can be strong or weak.

dimer: It refers to the composite molecule formed by the binding of two molecules (see homo and heterodimers).

dimethyl sulphoxide; Dimethyl sulfoxide (DMSO): It is the chemical compound with the formula $(CH_3)_2SO$. This colourless liquid is an important polar aprotic solvent that dissolves both polar and nonpolar compounds and is miscible in a wide range of organic solvents as well as water. It has a distinctive property of penetrating the skin very readily, so that one can taste it soon after it comes into contact with the skin. Its taste has been described as oyster- or garlic-like. Other reported side effects include stomach upset, sensitivity to light, visual disturbances, and headache. Skin irritation can develop at the site where DMSO is applied topically.

dimorphism: The existence of two different types of individuals within a species. An obvious example is the sexual dimorphism in certain animals.

dioxin: It refers to a highly toxic chlorinated hydrocarbon used in herbicides and produced by industrial pollution.

diploid: The status of having two complete sets of chromosomes, most commonly one set of paternal origin and the other of maternal origin. A cell which contains two sets of chromosomes is called a diploid cell.

diplonema: It is the chromosome at diplotene stage in which one pair of sister chromatids begins to separate from the other pair, i.e., the centromeres begin to disjoin.

Diplophase: Phase with 2n chromosomes.

direct embryogenesis: In plant tissue culture, the formation of embryoids directly in plant culture, without an intervening callus phase, on the surface of zygotic or somatic embryos or on explants.

direct organogenesis: In plant tissue culture, formation of organs directly on the surface of cultured explants without involving callus formation.

direct relationship: A relationship between two variables, in which as the value of independent variables increases, the value of the dependable variable also increases.

direct repeat: Direct repeats are nucleotide sequences present in multiple copies in the genome. There are several types of repeated sequences. Interspersed (or dispersed) DNA repeats (interspersed repetitive sequences) are copies of transposable elements interspersed throughout the genome. Flanking (or terminal) repeats (terminal repeat sequences) are sequences that are repeated on both ends of a sequence, for example, the long terminal repeats (LTRs) on retroviruses. Direct terminal repeats are in the same direction and inverted terminal repeats are opposite to each other in direction. Tandem repeats (tandem repeat sequences) are repeated copies which lie adjacent to each other. These can also be direct or inverted repeats. The ribosomal RNA and transfer RNA genes belong to the class of middle repetitive DNA.

directed mutagenesis: It is also known as directed mutation or adaptive mutation. It is a hypothesis that organisms can respond to environmental stress or advantages through directing mutations to certain genes or areas of the genome. It has long been observed in nature, with a classic example being sex change, with animals intentionally responding to their environmental conditions by changing their DNA between male and female for survival advantage.

directional cloning: DNA insert and vector molecules are digested with two different restriction enzymes to create noncomplementary sticky ends at either end of each restriction fragment. This allows the insert to be ligated to the vector in a specific orientation and prevent the vector from self-ligation.

disaccharide: A carbohydrate consisting of two linked sugar units.

disarming: Process of deleting sequences from a plasmid or virus genes that are cytotoxic or tumour inducing.

discontinuous variable: Variables which have only certain fixed numerical values with no intermediate values possible in between.

discrete data: Data that do not progress from one class to the next without a break, i.e., where classes represent distinct categories or counts and may be represented by whole numbers.

discrete probability distribution: A probability distribution in which the variable is allowed to take only a limited number of values.

discrete random variable: A discrete random variable is one that is allowed to take on only a limited number of values.

disease resistance: There are a number of lines of defence against pests (animals that cause damage to the plants we grow) and diseases in the organic garden, principle among these being the practice of good husbandry, creating healthy soil and ensuring high standards of garden hygiene. But no matter how diverse and healthy the garden ecosystem may be, there will always be a degree of disease and pest presence. In many ways, some level of pathogen population in the garden cannot only be acceptable but also desirable, as they indicate a generally healthful and diverse environment, and add to the overall robustness of the system as an immunity to such detrimental influences will build up, particularly in a balanced polycultural regime. Indeed, most of the plants we grow will tend to be selected because they are trouble free, and those that are more susceptible to attack will have fallen by the wayside over time. However, most farmers find it unacceptable that the food crops they grow are damaged by pests.

disease-free: Disease-free denotes the chances of staying free of disease after a particular treatment for a group of individuals suffering from a cancer. It is the percentage of individuals in the group who are likely to be free of disease after a specified duration of time. Disease-free survival rates are an indication of how effective a particular treatment is. Very often, two treatment strategies are compared on the basis of the disease-free survival that is achieved in similar groups of patients.

disease-indexing: Disease-indexed organisms have been assayed for the presence of known diseases according to standard testing procedures.

disinfectants: They are chemicals that are used to destroy microorganisms that are living on the objects of human use. Disinfectants are different from *antibiotics* that destroy microorganisms within the body, and from *antiseptics*, which destroy microorganisms on living tissue.

disinfestation: Complete elimination of microorganisms from the place of research or living. It also refers to the elimination of a threat of infestation by vermin, rodents, lice, or other noxious organisms.

disjunction: Separation of homologous chromosomes during anaphase.

disomy: The presence of a pair of specific chromosome.

dispersion: The scatter or variability in a set of data.

dissecting microscope: A microscope with a low magnifying power under which minor dissections like opening of flowers to study their parts is carried out.

dissection: Separation of a tissue by cutting for analysis or observation.

dissolved organic matter (DOM) or (dissolved organic carbon): It refers to decomposing carbon compounds in water. It can be natural or artificial.

distillation: It is a method of separating mixtures based on differences in their volatilities in a boiling liquid mixture. Distillation is a unit operation, or a physical separation process, and not a chemical reaction. Commercially, distillation has a number of uses. It is used to separate crude oil into more fractions for specific uses such as transport, power generation and heating. Water is distilled to remove impurities, such as salt from seawater. Air is distilled to separate its components, notably oxygen, nitrogen, and argon, for industrial use. Distillation of fermented solutions has been used since ancient times to produce distilled beverages with a higher alcohol content. The premises where distillation is carried out, especially distillation of alcohol, are known as a distillery.

disulphide bond (disulphide bridge): A chemical bond that stabilizes the three-dimensional structure of proteins, and hence the protein's normal function. They form between cysteine residues in the same or different peptide molecules.

disulphide bond: It refers to the covalent link formed between the sulphur atoms of two different cysteine residues in a protein. Important in maintaining the folded structure of a protein, and also for linking different proteins in a complex.

divergent evolution: It refers to a phenomenon, when two populations become more and more dissimilar, usually as a result of different environmental pressures.

divided bar diagram: In a divided bar diagram, the frequency is divided into different components, and such a representation is called a divided bar diagram.

dizygotic twins: A pair of individuals that share the same uterus at the same time, but arise from separate and independent fertilization of two ova.

DNA (deoxyribonucleic acid): It is one of the two forms nucleic acids that contain the genetic instructions used in the development and functioning of all known living organisms and some viruses. The main role of DNA molecules is the long-term storage of information. The DNA segments that carry this genetic information are called genes, but other DNA sequences have structural purposes, or are involved in regulating the use of this genetic information. Chemically, DNA consists of two long polymers of simple units called nucleotides, with backbones made of sugars and phosphate groups joined by ester bonds. These two strands run in opposite directions to each other and are therefore anti-parallel. The DNA chain is 22 to 26 Å wide (2.2 to 2.6 nanometres), and one nucleotide unit is 3.3 Å (0.33 nm) long. Although each individual repeating unit is very small, DNA polymers can be very large molecules containing millions of nucleotides.

DNA bank: It is a repository of DNA, usually used for research. The NIAS DNA Bank, for example, collects the DNA of agricultural organisms, such as rice and fish, for scientific research. Most DNA provided by DNA banks is used to develop more productive or environment-friendly agricultural species. Some DNA banks also store the DNA of rare or endangered species to ensure their survival.

DNA chip (DNA microarray): It is a multiplex technology used in molecular biology and medicine. It consists of an arrayed series of thousands of microscopic spots of DNA oligonucleotides, called features, each containing picomoles of a specific DNA sequence. This can be a short section of a gene or other DNA element that are used as probes to hybridize a cDNA or cRNA sample (called target) under high-stringency conditions. Probe-target hybridization is usually detected and quantified by the detection of fluorophore-, silver-, or chemiluminescence-labelled targets to determine relative abundance of nucleic acid sequences in the target. In standard microarrays, the probes are attached to a solid surface by a covalent bond to a chemical matrix (via epoxy-silane, amino-silane, lysine, polyacrylamide or others). The solid surface can be glass or a silicon chip, in which case they are commonly known as *gene chip* or colloquially *Affy chip* when an Affymetrix chip is used. Other microarray platforms, such as Illumina, use microscopic beads, instead of the large solid support. DNA arrays are different from other types of microarray only in that they either measure DNA or use DNA as part of its detection system. DNA microarrays can be used to measure changes in expression levels, to detect single nucleotide polymorphisms (SNPs), in genotyping or in resequencing mutant genomes. Microarrays also differ in fabrication, workings, accuracy, efficiency, and cost. Additional factors for microarray experiments are the experimental design and the methods of analyzing the data.

DNA cloning (gene cloning): Gene cloning is the technique of making copies, or clones, of a single gene. Once a gene is identified, clones can be made and used in different areas like, medical, industrial or research. Genetic engineering is the process of cloning genes into new organisms, or altering the DNA sequence to change the protein product. Genetic engineering depends on our ability to perform the following essential procedures. In the cloning process, the DNA is removed from cells, manipulations of the DNA are carried out in a test-tube, and the DNA is subsequently put back into cells. Because *E. coli* is so well characterized, it is usually the cell of choice for manipulating DNA molecules. Once the appropriate combination of vector and cloned DNA or construct has been made in *E. coli*, the construct can be put into other cell types.

DNA construct: A DNA molecule inserted into a cloning vector.

DNA diagnosis: The use of DNA polymorphisms to detect the presence of a disease gene.

DNA fingerprinting (DNA profiling or DNA testing or DNA typing): Using the specific sequences of DNA that are unique to every person, his or her identity could be determined solely by the sequence of their base pairs. However, because there are so many millions of base pairs, the task would be very time-consuming. Instead, scientists are able to use a shorter method, because of repeating patterns in DNA. These patterns do not, however, give an individual "fingerprint," but they are able to determine whether two DNA samples are from the same person, related people, or non-related people. Scientists use a small number of sequences of DNA that are known to vary among individuals a great deal, and analyze those to get a certain probability of a match. This technique is employed by forensic scientists to assist in the identification of individuals on the basis of their respective DNA profiles.

DNA helicase (gyrase): An enzyme that catalyses the unwinding of complementary strands of a DNA double helix.

DNA hybridization: Hybridization is the process of establishing a non-covalent, sequence-specific interaction between two or more complementary strands of nucleic acids into a single hybrid, which in the case of two strands is referred to as a duplex. Oligonucleotides, DNA or RNA, will bind to their complement under normal conditions, so two perfectly complementary strands will bind to each other readily. In order to reduce the diversity and obtain the most energetically preferred hybrids, a technique called annealing is used in the laboratory practice. However, due to the different molecular geometries of the nucleotides, a single inconsistency between the two strands will make binding between them less energetically favourable. Measuring the effects of base incompatibility by quantifying the rate at which two strands anneal can provide information as to the similarity in base sequence between the two strands being annealed. The hybrids may be dissociated by thermal denaturation, also referred to as melting. Here, the solution of hybrids is heated to break the hydrogen bonds between nucleic bases, after which the two strands separate. In the absence of external negative factors, the processes of hybridization and melting may be repeated in succession indefinitely long, which lays the ground for polymerase chain reaction. Most commonly, the pairs of nucleic bases A=T and G≡C are formed, of which the latter is more stable.

DNA ligase: They close nicks in the phosphodiester backbone of DNA. Biologically, DNA ligases are essential for the joining of Okazaki fragments during replication, and for completing short-patch DNA synthesis occurring in DNA repair process. There are two classes of DNA ligases. The first uses NAD+ as a cofactor and is found only in bacteria. The second uses ATP as a cofactor and is found in eukaryotes, viruses and bacteriophages. The smallest known ATP-dependent DNA ligase is the one from the bacteriophage T7 (at 41KdA). Eukaryotic DNA ligases may be much larger (human DNA ligase I is > 100KDA), but they all appear to share some common sequences and probably structural motifs.

DNA polymerase: An enzyme that catalyses the synthesis of double-stranded DNA using single-stranded DNA as a template. Based on sequence homology, DNA polymerases can be further subdivided into seven different families: A, B, C,

D, X, Y, and RT. Error correction is a property of some, but not all, DNA polymerases. This process corrects mistakes in newly-synthesized DNA. When an incorrect base pair is recognized, DNA polymerase reverses its direction by one base pair of DNA. The 3'->5' exonuclease activity of the enzyme allows the incorrect base pair to be excised (this activity is known as *proofreading*). Following base excision, the polymerase can re-insert the correct base and replication can continue.

DNA primase: An enzyme that catalyses the synthesis of short strands of RNA that initiate the synthesis of DNA strands.

DNA probe: A labelled segment of DNA that is able to detect a specific DNA sequence in a mixture of sequences. If the tagged sequence is complementary to any one in the mixture, the two sequences will form a double helix.

DNA profiles: These are encrypted sets of numbers that reflect a person's DNA makeup, which can also be used as the person's identifier. DNA profiling should not be confused with full genome sequencing. Although 99.9% of human DNA sequences are the same in every person, enough DNA is different to distinguish one individual from another. DNA profiling uses repetitive ("repeat") sequences that are highly variable, called variable number tandem repeats (VNTR). VNTRs loci are very similar between closely related humans, but so variable that unrelated individuals are extremely unlikely to have the same VNTRs. The DNA profiling technique was first reported in 1985 by Sir Alec Jeffreys at the University of Leicester in England, and is now the basis of several national DNA databases.

DNA repair enzymes: Enzymes that catalyse the repair of DNA.

DNA repair: A variety of mechanisms that repair errors occurring during DNA replication.

DNA replication: DNA replication is the process whereby an entire double-stranded DNA is copied to produce a second, identical DNA double helix. In a cell, DNA replication begins at specific locations in the genome, called "origins". Unwinding of DNA at the origin, and synthesis of new strands, forms a replication fork. In addition to DNA polymerase, the enzyme that synthesizes the new DNA by adding nucleotides matched to the template strand, a number of other proteins are also associated with the fork and assist in the initiation and continuation of DNA synthesis.

DNA sequencing: Procedures for determining the nucleotide sequence of a DNA fragment. The term DNA sequencing refers to sequencing methods used for determining the order of the nucleotide bases—adenine, guanine, cytosine, and thymine—in a molecule of DNA. Knowledge of DNA sequences has become indispensable for basic biological research, other research branches utilizing DNA sequencing, and in numerous applied fields such as diagnostic, biotechnology or forensic biology. The advent of DNA sequencing has significantly accelerated biological research and discovery. The rapid speed of sequencing attained with modern DNA sequencing technology has been instrumental in the sequencing of human genome in the Human Genome Project. Related projects, often by scientific collaboration across the continents, have generated the complete DNA sequences of many animals, plants, and microbial genomes. The first DNA sequences were obtained by academic researchers using laborious methods based on 2-dimensional chromatography in the early 1970s. Following the development of dye-based sequencing methods with automated analysis, DNA sequencing has become easier and orders of magnitude faster.

DNA topoisomerase: An enzyme that catalyses the introduction or removal of supercoils in DNA.

DNase (deoxyribonuclease): An enzyme that catalyses the cleavage of DNA. DNase I is a digestive enzyme secreted by the pancreas that degrades DNA into shorter nucleotide fragments. Many other endonucleases and exonucleases are involved in DNA repair and replication.

Dollo's law: It refers to the law according to which evolution never reverses itself.

Dolly: The first mammal created by cloning from the somatic cell of an adult animal.

domain (protein): It refers to the region of special biological interest within a single protein sequence. However, a domain may also be defined as a region within the three-dimensional structure of a protein that may encompass regions of several distinct protein sequences that accomplishes a specific function. A domain class is a group of domains that share a common set of well-defined properties or characteristics.

domain: A segment of a protein that has a discrete function or conformation. At the protein level, a domain can be as small as a few amino acid residues or as large as half of the entire protein.

dominant marker selection: Selection of cells via a gene encoding a product that enables only the cells that carry the gene to grow under particular conditions. For example, plant and animal cells that express the introduced *Neor* gene are resistant to the compound G418, while cells that do not carry the *Neor* gene are killed by G418.

dominant oncogene: A gene that stimulates cell proliferation and contributes to oncogenesis even when present in a single copy.

dominant selectable marker gene: A gene that allows the host cell to survive under conditions where it would otherwise die.

dormancy: It is an inactive period in the life of an animal or plant during which the growth slows down or ceases completely. During this state, the physiological changes help the organism survive adverse environmental conditions.

double fertilization: It is a process in which two male nuclei fuse separately with different female nuclei in the embryo sac. The first one fuses with the egg cell to form the zygote, and the second one fuses with the two polar nuclei to form a triploid nucleus that develops into an endosperm.

double helix: The coiling of antiparallel strands of DNA molecule, resembling a spiral staircase.

double recessive: An organism homozygous for a recessive allele.

double-stranded complementary DNA (dscDNA): A double-strand DNA molecule created from a cDNA template.

downcycling: It refers to the process of converting waste or discarded products into new products with reduced quality and lower functionality. It is actually a recycling process when the resultant material is below the quality of the original source. An example of this is the conversion of plastics into lower-quality plastics.

downstream: Downstream and upstream both refer to a relative position in DNA or RNA. Each strand of DNA or RNA has a 5' end and a 3' end, so named for the carbons on the deoxyribose (or ribose) ring. Relative to the position on the strand, downstream is the region towards the 3' end of the strand. Since DNA strands run in opposite directions, downstream on one strand is upstream on the other strand.

drug delivery: Method by which a therapeutic agent is delivered to its site of action.

drug discovery cycle: It refers to the cycle of events required to develop a new drug. Typically this involves research, preclinical testing and clinical development, and can take from 5 to 12 years.

drug: It refers to the agent that affects a biological process. Specifically, a molecule whose molecular structure can be correlated with its pharmacological activity.

dry weight: The moisture-free weight of tissue obtained by drying.

dual culture: A culture made of a plant tissue and one organism

dust dome: It refers to brown or gray dome of air particles and pollutants trapped by an urban heat island above a city.

dykes: It refers to vertical veins of igneous rock formed by magma cooling in breaks and fractures.

dynamic metamorphism: It refers to metamorphism that changes a rock's shape without changing it chemically. Sometimes associated with mountain-building.

dystrophic: It refers to toxic habitats deficient in nutrients.

E

ecesis: It refers to the establishment of a plant or animal species in a new environment.

ecological equivalent: It refers to species that live far apart but in similar niches and ecosystems.

ecological footprint: It refers to the amount of biologically productive land and sea area necessary to supply the resources a human population consumes, and to assimilate associated waste. The current estimates suggest that, as a global community, we currently need about 1.3 planets to meet our average consumption levels.

ecotherapy: It refers to the earth-based healing practices.

ecotone: It refers to the transition zone between adjacent biotic communities, often with unique nutrients and ecological relationships.

ecotopia: It refers to a vision of an ecologically friendly society.

ecotype: It refers to a genetically differentiated subpopulation evolved to remain within its habitat.

EDTA (ethylene-diamine tetra-acetic acid: It is usually used as a chelating agent in the preparation of nutrient media. EDTA is a polyamino carboxylic acid with the formula $[CH_2N(CH_2CO_2H)_2]_2$ and is a colourless, water-soluble solid widely used to dissolve limescale (hard and chalky deposit usually found in hot-water boilers and the inside of inadequately maintained hot-water central heating systems, it is made up of calcium salts). Its usefulness ariscs because of its role as a chelating agent, i.e., its ability to sequester metal ions such as Ca^{2+} and Fe^{3+}. After being bound by EDTA, metal ions remain in solution but exhibit diminished reactivity. EDTA is produced as several salts, notably disodium EDTA and calcium disodium EDTA.

effector molecule: A molecule that influences the behaviour of a regulatory molecule.

effluent: It refers to a substance (usually toxic) entering an environment from a point source (a confirmable location), e.g., wastewater from a manufacturing plant (effluence from affluence).

EIA: It is the abbreviation for enzyme immunoassay and is another term used for ELISA.

elastin: A fibrous protein of the yellow elastic fibres of connective tissue.

electro-blotting: It is a method in molecular biology/biochemistry/immunogenetics in which proteins or nucleic acids are transferred onto a membrane by using PVDF, rather than nitrocellulose, after gel electrophoresis. The protein or nucleic acid can then be further analyzed using probes such as specific antibodies or stains. This method can be used with all polyacrylamide and agarose gels. An alternative technique for transferring proteins from a gel is capillary blotting.

electrochemical sensor: A type of biosensor in which a biological process is harnessed to an electrical sensor system.

electromagnetic radiation: Electromagnetic waves, including UV, X and gamma radiation. Electromagnetic radiation is used to produce mutant cells or in tissue culture.

electromagnetic spectrum: The range of wavelengths over which electromagnetic radiation extends.

electron microscope: A microscope that uses an electron beam focused by magnetic lenses.

Electronic Northerns: It refers to the use of an electronic database of cDNA sequences (or probes derived from them) in order to measure the relative levels of mRNAs expressed in different cells or tissues. An example of the use of an electronic Northern might be to identify the differences in the genes expressed in prostate cancer and those in benign prostate hyperplasia, by subtracting the database of one from the other and seeing which cDNAs remain.

electrophoresis: A technique that separates charged molecules such as DNA, RNA or protein on the basis of relative migration in an appropriate matrix subjected to an electric field.

electroporation: It is a significant increase in the electrical conductivity and permeability of the cell plasma membrane caused by an externally applied electrical field. It is usually used in molecular biology as a way of introducing some substance into a cell, such as loading it with a molecular probe, a drug that can change the cell's function, or a piece of coding DNA. In molecular biology, the process of electroporation is often used for the transformation of bacteria, yeast, and plant protoplasts. In addition to the lipid membranes, bacteria also have cell walls which are different from the lipid membranes and are made of peptidoglycan and its derivatives. However, the walls are naturally porous and only act as stiff shells that protect bacteria from severe environmental impacts. If bacteria and plasmids are mixed together, the plasmids can be transferred into the cell after electroporation. Several hundred volts across a distance of several millimetres are typically used in this process. Afterwards, the cells have to be handled carefully until they have had a chance to divide producing new cells that contain reproduced plasmids. This process is approximately ten times as effective as chemical transformation.

ELISA (enzyme-linked immunosorbent assay): It is a biochemical technique used mainly in immunology to detect the presence of an antibody or an antigen in a sample. The ELISA has been used as a diagnostic tool in medicine and plant pathology, as well as a quality control check in various industries. In simple terms, in ELISA an unknown amount of antigen is affixed to a surface, and then a specific antibody is washed over the surface so that it can bind to the antigen. This antibody is linked to an enzyme, and in the final step a substance is added so that the enzyme can convert to some detectable signal. Thus, in the case of fluorescence ELISA, when a light of appropriate wavelength is shone upon the sample, any antigen/antibody complex will fluoresce so that the amount of antigen in the sample can be inferred through the magnitude of fluorescence.

elongation factors: These are a set of proteins that facilitate the events of translational elongation, the steps in protein synthesis from the formation of the first peptide bond to the formation of the last one. Elongation is the most rapid step in translation. In prokaryotes it proceeds at a rate of 15 to 20 amino acids added per second (about 60 nucleotides per second). In eukaryotes the rate is about two amino acids per second. Elongation factors play a role in orchestrating the events of this process, and ensuring 99.99% accuracy of translation at this speed.

eluviation: It refers to a movement of humus, chemical substances, and mineral particles from the upper layers of a soil to lower layers by the downward movement of water through the soil profile.

embryo culture: The culture of isolated embryos on nutrient media. It is one of the major techniques used for embryo rescue in plant tissue culture.

embryo sexing: Sexing of embryos before transfer and implanting has great potential for the livestock industry. Manipulating the sex of offspring has been a dream of the cattle industry for decades. A number of approaches to the sexing of semen have been attempted, and several have been reported as successful. However, the only method of semen sexing that has shown any promise has been the sorting of spermatozoa according to the DNA content, by means of flow cytometry. The techniques involving the determination of the sex of an embryo, usually by PCR.

embryo splitting: The splitting of young embryos into several sections. It is in practice to produce many embryos from a single embryo, when more number of embryos are required.

embryogenesis: It refers to the formation and development of an embryo, especially in plant tissue culture by inducing with hormones.

embryoid: It is an embryo-like body produced in plant tissue culture from which a complete plant can be produced.

embryonic stem cells: Stem cells derived, which are embryonic in origin.

emissivity: It refers to the quantity of heat flowing from the earth back into space. Emissivity and albedo help balance the global temperature.

encapsidation: The process by which the nucleic acid of virus gets enclosed in a capsid.

encapsulating agents: Substances which are used to coat around embryoids, tissues enzymes, or bacterium.

encapsulation: It is a method of producing artificial seeds by coating embryoids or tissues with nutrients and other suitable substances. The term is also used for the technique in which cells are immobilized for use in a bioreactor.

end-labelling: The introduction of a radioactive atom at the end of a DNA or RNA molecule.

endocrine disruptor: It refers to a pesticide chemical that interferes with hormones. It is suspected to cause vaginal cancer, immune system deficiencies, and birth defects.

endocytosis: The process by which materials enter a cell without passing through the cell membrane.

endogenote: The part of the bacterial chromosome that is homologous to a genome fragment transferred from the donor to the recipient cell.

endogenous: Anything that is developed or added from within the cell or organism.

endomitosis: Replication of chromosomes without separation. This will lead to endoduplication of chromosomes. The number of chromosomes increases resulting in polyploidy. It naturally occurs in the polytene chromosomes of *Drosophila*. The process can also be induced using chemicals.

endonuclease: These are enzymes that cleave the phosphodiester bond within a polynucleotide chain, in contrast to exonucleases, which cleave phosphodiester bonds at the end of a polynucleotide chain. *Restriction endonucleases* (Restriction Enzymes) cleave DNA at specific sites, and are divided into three categories—Type I, Type II, and Type III, according to their mechanism of action. These enzymes are often used in genetic engineering to make recombinant DNA for introduction into bacterial, plant, or animal cells. Restriction endonucleases (ENases) are products of bacteria, and can be used to map a piece of DNA. Some endonucleases have actions on RNA, such as the Dicer enzyme, which initiates the formation of RNA-induced silencing complexes. These may also be termed endoribonucleases.

endophyte: An organism that lives inside a plant.

endopolyploidy: The process of producing polyploids by the multiplication of chromosomes, without nuclear division.

endoprotease: An enzyme that cleaves the peptide bonds between amino acids within a protein.

endotoxin: A toxin produced by Gram-negative bacteria that elicits an inflammatory response.

end-product inhibition: The inhibition of an enzyme once the end product is formed with the help of the metabolite.

enhancer element (enhancer sequence): A DNA sequence that increases the transcription of a eukaryotic gene when they are both on the same DNA molecule.

enhancers: It refers to the DNA sequences that can greatly increase the transcription rates of genes even though they may be far upstream or downstream from the promoter they stimulate.

enterotoxin: A bacterial toxin that is released into the intestine and causes cramps, diarrhoea and nausea.

entropy: It refers to disorder or unproductive energy in a given system. It tends to increase over time.

enucleated ovum: Egg without the nucleus.

enzyme classification: The names of enzymes are often derived from their substrate or the chemical reactions they catalyze, with the word ending in ***-ase***. The International Union of Biochemistry and Molecular Biology have developed a nomenclature for enzymes, the **EC numbers**; each enzyme is described by a sequence of four numbers preceded by "EC". The first number broadly classifies the enzyme based on its mechanism: *Oxidoreductases*: catalyze oxidation/reduction reactions; *Transferases*: transfer a functional group (e.g., a methyl or phosphate group); *Hydrolases*: catalyze the hydrolysis of various bonds; *Lyases*: cleave various bonds by means other than hydrolysis and oxidation; *Isomerases*: catalyze isomerization changes within a single molecule; *Ligases*: join two molecules with covalent bonds.

enzyme commission (EC) number: Systematic name and number which identify an enzyme in technical literature, e.g., DNA gyrase, DNA topoisomerase II. (**EC** 5.99).

enzyme electrode: A type of biosensor in which an enzyme is immobilized onto the surface of an electrode. When the enzyme catalyses its reaction, electrons are transferred from the reactant to the electrode, and a current is generated.

enzyme stabilization: A method of stabilizing enzymes by binding antibodies to them.

enzyme: These are proteins that catalyze (i.e., increase the rate of) chemical reactions. In enzymatic reactions, the molecules at the beginning of the process are called substrates, and the enzyme converts them into different molecules, called the products. Almost all processes in a biological cell need enzymes to occur at significant rates. Since enzymes are selective for their substrates and speed up only a few reactions from among many possibilities, the set of enzymes made in a cell determines which metabolic pathways occur in that cell. Like all catalysts, enzymes work by lowering the activation energy (E_a or $\Delta G^{\ddagger}$) of a reaction, thus dramatically increasing the rate of reaction. Most enzyme reaction rates are millions of times faster than those of comparable uncatalyzed reactions. As with all catalysts, enzymes are not consumed by the reactions they catalyze, nor do they alter the equilibrium of these reactions. However, enzymes do differ from most other catalysts by being much more specific. Enzymes are known to catalyze about 4000 biochemical reactions. A few RNA molecules called ribozymes also catalyze reactions, with an important example being some parts of the ribosome. Synthetic molecules called artificial enzymes also display enzyme-like catalysis.

epifauna: It refers to animals that live on a surface.

epigenesis: The developmental process whereby each successive stage of normal development is built up on the foundations created by the preceding stages of development.

epigenetic variation: Non-hereditary and reversible variation.

epigenomics: It refers to the study of complex expression networks or linkages both spatially (within the body) and temporally (at different times in development).

episome: An extrachromosomal, independent and self replicating genetic element, which is found in addition to the chromosome of the organisms like bacteria. Plasmids and F factors are episomes.

epistasis: Interaction between the genes at different loci.

epitope: A specific chemical domain on an antigen that is recognized by an antibody and stimulates the production of specific immunoglobulins.

equilibrium constant: It refers to the value that describes the equilibrium state of the reversible reaction between two molecular species.

equilibrium density gradient centrifugation: A type of centrifugation used to separate macro-molecules based on their density.

equilibrium success strategy: It refers to species that survive through low death rate, relatively little colonization or expansion, moderately frequent reproduction cycles, and other steady-state patterns of adaptation.

equimolar: Identical molar concentrations.

***Escherichia coli*:** A bacterium inhabiting the colon of many species. It is widely used in biotechnology, especially, as a host for molecular cloning experiments.

essential amino acid: An amino acid that cannot be synthesized by animals.

essential nutrient: A nutrient required by living organisms to ensure normal growth, development and maintenance.

EST: It refers to the expressed sequence tag. They may be used to identify gene transcripts, and are instrumental in gene discovery and gene sequence determination.

established culture: An aseptically developed culture produced by inducing an explant in vitro.

estimation equation: A statistical formula that relates the unknown variable to the known variable in regression analysis.

ethanol (ethyl alcohol): A common surface sterilant used in plant tissues to sterilize glassware, utensils and working surfaces. It is used in 70% concentration (v/v) for disinfecting and 95% (v/v) when flaming tools. It is also used to dissolve water-insoluble additions to culture media.

ethephon (2-chloroethyl) phosphonic acid: It is a synthetic compound commonly used to treat cultured cells or unripened fruit with ethylene.

ethidium bromide: A fluorescent dye used to stain DNA and RNA.

ethylene: A plant growth regulator generally used for fruit ripening and abscission of plant parts.

etiolation: It is caused by the lack of chlorophyll, and is typical of plants growing under low light intensity or complete darkness. It is characterized by an abnormal increase in stem elongation, accompanied by poor or absent leaf development. It can also be caused by disease.

euchromatin: It is a lightly packed form of chromatin that is rich in gene concentration, and is often under active transcription. It is stained less intensely by chromatin dyes during the process of chromosome staining.

eugenics: It is the "applied science or the biosocial movement which advocates the use of practices aimed at improving the genetic composition of a population," usually referring to human populations. Eugenics was widely popular in the early decades of the 20th century, but has fallen into disfavour after having become associated with Nazi Germany and with the discovery of molecular evolution. Since the postwar period, both the public and the scientific communities have associated eugenics with Nazi abuses, such as enforced racial hygiene, human experimentation, and the extermination of "undesired" population groups.

eukaryote: A higher organism with a true nucleus having chromosomes within a nuclear membrane. They also possess membrane-bound functional organelles, such as mitochondria and chloroplasts in the cytoplasm of their cells.

euploid: An organism or cell having a complete set or its multiples of chromosomes.

eutrophication: It refers to a gradual nutrient enrichment that increases organic matter production, as with a lake that absorbs waves of nutrient run-offs.

evapotranspiration (transpiration): It refers to the evaporation of water off leaf surfaces. A natural event, but can cause plants to lose too much water in dry times.

event: One of the possible outcomes which results from conducting an experiment.

evolution: The process by which the present diversity of plant and animal life arose from the earliest organisms. It is caused by the accumulation of heritable changes that constantly occur in nature. Evolution is by chance and not by choice.

excinuclease: The endonuclease-containing protein complex that excises a segment of damaged DNA during excision repair.

excision repair: DNA repair processes that involve the removal of a damaged or incorrect segment of DNA and its replacement by the synthesis of a new strand using the complementary strand of DNA as template.

excision: Cutting out or removing of tissue or organ in plant culture.

exit site (E site): It refers to exit site on the ribosome, and it is at this site the ribosome carrying amino acid exits during the translation process of protein synthesis.

exogenote: Chromosomal fragment homologous to an endogenote and donated to a merozygote.

exogenous: Something that is produced outside the organism due to external causes.

exon amplification: A procedure that is used to amplify exons.

exon: Eukaryotic gene is made up two types of regions: exons and introns. Exon is transcribed as part of the primary transcript and is retained, after processing, with other exons, to form a functional mRNA molecule.

exonuclease: An enzyme that digests DNA or RNA, beginning at the ends of strands.

exopolysaccharide: A high molecular weight polymer secreted by a microorganism into the surrounding environment.

exotic species: It refers to those not native to an ecosystem.

exotoxin: A toxin released by a bacterium into the medium in which it grows.

expected frequency: The frequencies we would expect to see in a contingency table or frequency distribution if the null hypothesis is true.

experiment: The activity that results in an event.

explant: A plant part aseptically excised and prepared for plant tissue culture.

exponential phase: Growth of plants both in nature and culture occurs in different stages. Exponential phase is a stage in which cells undergo their maximum rate of cell division. The exponential phase follows the lag phase, and precedes the linear growth phase.

expressed sequence tag (EST): Short length of DNA sequences representing part of an expressed gene. It is actually a short cDNA sequence.

expression library: A population of different DNA molecules encoding peptides that have been cloned into one kind of expression vector.

expression profile: It refers to the level and duration of expression of one or more genes, selected from a particular cell or tissue type, generally obtained by a variety of high-throughput methods, such as sample sequencing, serial analysis, or microarray-based detection.

expression vector: A cloning vector that is usually a plasmid or virus designed for protein expression in cells. The vector is used to introduce a specific gene into a target cell, and can commandeer the cell's mechanism for protein synthesis to produce the protein encoded by the gene. The cloned gene is put under the control of a promoter sequence for the initiation of transcription, and often also has a transcription termination sequence at its end.

expression vector: It refers to the cloning vector that is engineered to allow the expression of protein from a cDNA. The expression vector provides an appropriate promoter and restriction sites that allow insertion of cDNA.

expressivity: Degree of expression of a trait controlled by a gene.

extrachromosomal inheritance (cytoplasmic inheritance): Inheritance of traits through DNA that is not connected with the chromosomes but rather with organelles in the cell.

extrachromosomes: The genetic elements found in addition to main chromosome(s) of a cell, e.g., rDNA molecules of yeast plasmid, dsRNA molecules in fungi, mitochondrial DNA, and chloroplast DNA.

F

F factor: A bacterial episome that confers the ability to function as a genetic donor in conjugation; the fertility factor in bacteria.

F: It stands for filial in genetics and is used as F1 for first filial generation. Likewise it is used as F2, second filial generation.

FACS: It stands for Facial Action Coding System and is a system that taxonomizes human facial expressions.

factorial mating: A mating scheme in which each male parent is mated with each female parent. Made possible in animals by means of in vitro embryo production.

FAD (flavin adenine dinucleotide): A coenzyme important in various biochemical reactions. It comprises a phosphorylated vitamin B2 (riboflavin) molecule linked to nucleotide adenine monophosphate (AMP).

FDA or USFDA (The Food and Drug Administration): It is an agency of the United States Department of Health and Human Services, one of the United

States federal executive departments, responsible for protecting and promoting public health through the regulation and supervision of food safety, tobacco products, dietary supplements, prescription and over-the-counter pharmaceutical drugs (medications), vaccines, biopharmaceuticals, blood transfusions, medical devices, electromagnetic radiation emitting devices (ERED), veterinary products, and cosmetics.

F-distribution: A type of distribution differentiated by two parameters (df–numerator, df-denumerator), used primarily to test hypothesis regarding variances.

fed-batch fermentation: Culture of cells or microorganisms where nutrients are added periodically to the bioreactor.

feedback inhibition: The process by which the accumulated end product of a biochemical pathway stops the synthesis of that product. A late metabolite of a synthetic pathway regulates synthesis at an earlier step of the pathway.

fermentation substrates: Materials used as food for growing microorganisms. The fermentation substrates and the trace materials needed, together with chemicals added to make the fermentation easier, form the culture medium.

fermentation: The anaerobic breakdown of complex organic substances, especially carbohydrates like glucose by microorganisms. The process is energy-yielding. Fermentation is often misused to describe large-scale aerobic cell culture in specialized vessels (fermenters, bioreactors) for secondary product synthesis.

Feulgen's test: A histochemical test in which the distribution of DNA in the chromosomes of dividing cell nuclei can be observed. A tissue section is first treated with dilute hydrochloric acid to remove the purine bases of the DNA, thus exposing the aldehyde groups of the sugar deoxyribose. The section is then immersed in Schiff's reagent, which combines with the aldehyde groups to form a magenta-coloured compound.

filter sterilization: Process of sterilizing a liquid by passage through a filter with pores so small that they prevent the passage of microorganisms and microbial spores.

filtration: Separation of solids from liquids by using a porous material that allows only the passage of liquid or solids of a size smaller than the pore size of the filter. The material passing through the filter forms the filtrate.

fine particulate organic matter (FPOM): It refers to organic matter which has undergone an initial decomposition.

fingerprint: Fingerprint is a set of motifs used to predict the occurrence of similar motifs, in either an individual sequence or in a database. Fingerprints are refined by iterative scanning of a composite protein sequence database. A composite or multiple-motif fingerprint contains a number of aligned motifs taken from different parts of a multiple alignment. True family members are then easy to identify by virtue of possessing all elements of the fingerprint, while subfamily members may be identified by possessing only part of it.

fission: Asexual reproduction involving the division of a single-celled individual into two new single-celled individuals of equal size.

fixation: A method of preserving plant materials using chemicals to maintain them in a particular stage of cell division to be used later for cytological studies. Usually roots and flower buds are fixed in a solution containing acetic acid and alcohol.

flagellum: An organelle of locomotion in certain cells, spores, gametes or organisms. Flagella may be either whiplash-like or tinsel (feather)-shaped.

flaming: A technique used for heat sterilization of instruments. They are dipped in alcohol and then the alcohol on the instrument is ignited, thus heating the tool's surface.

flanking region: The DNA sequences extending either side of a specific sequence.

floccules: Microorganism aggregate or colloidal particle floating in or on a liquid. Contaminated liquid media are usually cloudy, illustrating this flocculation phenomenon. An agent that causes small particles to aggregate is called flocculant.

flow cytometry: The original name of flow cytometry technology was "pulse cytophotometry". Only 20 years later in 1988, at the Conference of the American Engineering Foundation in Pensacola, Florida, the name was changed to "flow cytometry", a term that quickly became popular. It is a technique for counting and examining microscopic particles, such as cells and chromosomes, by suspending them in a stream of fluid and passing them through an electronic detection apparatus. It allows simultaneous multi-parametric analysis of the physical and/or chemical characteristics of up to thousands of particles per second. Flow cytometry is routinely used in the diagnosis of health disorders, especially blood cancers, but has many other applications in both research and clinical practice. A common variation is to physically sort particles based on their properties, so as to purify populations of interest.

fluorescence in situ hybridization (FISH): This is a major method of physical mapping of cloned DNA fragments on chromosomes. It is the hybridization of cloned DNA to intact chromosomes, where the cloned DNA has been labelled with a fluorescent dye.

fluorescence-activated cell sorting (FACS): The use of laser beams to detect differences in fluorescence between different types of cells in a mixture, and the subsequent deflection of cells into separate bins corresponding to each type of cell in the mixture. One of the popular uses of this technology is in sperm sexing.

fluorescent probe: Probe whose response is based on the fluorescence intensity of individual cells or cell components.

fog (mist) propagation: It is a method of propagating plants, especially those grown in vitro by using fine particles of liquid suspended in the air, such as water in a fog chamber.

fold-back: The structure formed when a double-stranded DNA molecule containing an inverted repeat sequence is denatured and then allowed to re-anneal at low DNA concentrations. The repeated sequence permits the formation of a double-stranded region within each of the separated strands of the original molecule.

folded genome: The condensed intracellular state of the DNA in the nucleoid of a bacterium. The DNA is segregated into domains, and each domain is independently negatively supercoiled.

follicle: Any enclosing cluster of cells that protects and nourishes a cell or structure within. Thus, a follicle in the ovary contains a developing egg cell, while a hair follicle envelops the root of hair.

food processing enzyme: Enzyme used to control food texture, flavour, appearance and, to a certain extent, nutritional value. Amylases break down complex polysaccharides to simplex sugars; proteases tenderize meat proteins. Biotechnology can assist development of new food enzymes by finding or engineering enzymes, which fit better with the other processes that the food must undergo, like cooking or canning.

forced cloning: The insertion of foreign DNA into a cloning vector in a predetermined orientation.

foreign (exogenous) DNA: A DNA molecule that is incorporated into either a cloning vector or a chromosomal site.

formulation: In pharmaceutics, it is the process in which different chemical substances, including the active drug, are combined to produce a final medicinal product. Formulation studies involve developing a preparation of drug, which is both stable and acceptable to the patient. For orally taken drugs, this usually involves incorporating the drug into a tablet or a capsule. It is important to appreciate that a tablet contains a variety of other substances apart from the drug itself, and studies have to be carried out to ensure that the drug is compatible with other substances.

fouling: It refers to the accumulation of unwanted material on solid surfaces, most often in an aquatic environment. The fouling material can consist of either living organisms (biofouling) or a non-living substances (inorganic or organic). Fouling is usually distinguished from other surface-growth phenomena in that it occurs on the surface of a component, system or plant performing a defined and useful function, and the fouling process impedes or interferes with this function.

four-base cutter (four-base-pair-cutter or four-cutter): It is a type II restriction endonuclease that binds (and subsequently cleaves) DNA at sites containing a sequence of four nucleotide pairs uniquely recognised by that enzyme. Because any sequence of four bases occurs more frequently by chance than any sequence of six bases, four-base cutters cleave more frequently than six-base cutters. Thus, four-base cutters create smaller fragments than six-base cutters.

fractionation: It is a separation process in which a certain quantity of a mixture (solid, liquid, solute, suspension or isotope) is divided in a number of smaller quantities (fractions) in which the composition changes according to the gradient. Fractions are collected based on differences in a specific property of the individual components. A common trait in fractionation is the need to find an optimum between the amount of fractions collected and the desired purity in each fraction. Fractionation makes it possible to isolate more than two components in a mixture in a single run. This property sets it apart from other separation techniques. Fractionation is widely employed in many branches of science and technology. Mixtures of liquids and gases are separated by fractional distillation by difference in boiling point. Fractionation of components also takes place in column chromatography by a difference in affinity between stationary phase and mobile phase. In fractional crystallization and fractional freezing, chemical substances are fractionated based on the difference in solubility at a given temperature. In cell fractionation, cell components are separated by difference in mass.

fragment (restriction): A restriction fragment is a DNA fragment resulting from the cutting of a DNA strand by a restriction enzyme (restriction endonucleases), a process called restriction. Each restriction enzyme is highly specific, recognising a particular short DNA sequence, or restriction site, and cutting both DNA strands at specific points within this site. Most restriction sites are palindromic, (the sequence of nucleotides is the same on both strands when read in 5' to 3' direction), and are four to eight nucleotides long. Many cuts are made by one restriction enzyme because of the chance repetition of these sequences in a long DNA molecule, yielding a set of restriction fragments. A particular DNA molecule will always yield the same set of restriction fragments when exposed to same restriction enzyme. Restriction fragments can be analyzed using techniques such as gel electrophoresis or used in recombinant DNA technology.

frameshift mutation: It is a genetic mutation caused by indels, i.e., insertion or deletion of a number of nucleotides that is not evenly divisible by three from a DNA sequence. Due to the triplet nature of gene expression by codons, the insertion or deletion can disrupt the reading frame, or the grouping of the codons, resulting in a completely different translation from the original. A frameshift mutation causes the reading of all codons after the mutation to code for different amino acids (with a few exceptions due to redundancy). Furthermore, the stop codon ("UAA", "UGA" or "UAG") will not be read or a stop codon could be created at an earlier or later site. The protein being created could be abnormally short, abnormally long, and/or contain the wrong amino acids. It will most likely not be functional.

frameshift: It refers to the deletion, substitution, or duplication of one or more bases that cause the reading-frame of a structural gene to shift from the normal series of triplets.

F-ratio: A ratio used in the analysis of variance, among other tests, to compare the magnitude of two estimates of the population variance to determine if the two estimates are approximately equal; in ANOVA, the ratio of between–sample variance to within–sample variance is used.

free market: It refers to an economic system in which the wealthiest players are free to make as much money as they can regardless of consequences or catastrophes.

freeze-drying (lyophilization or cryodesiccation): It is a dehydration process used to preserve a perishable material or make the material more convenient for transport. Freeze-drying works by freezing the material and then reducing the surrounding pressure and adding enough heat to allow the frozen water in the material to sublime directly from the solid phase to gas.

frequency curve: A frequency polygon smoothed by adding classes and data points to a data set.

frequency dependence: It refers to an interspecific equivalent of density-dependence in which a species grows when its population is low compared to another species and stops growing when the population is high.

frequency distribution: An organised display of data that shows the number of observations from that data set that falls into each of a set of mutually exclusive classes.

frequency polygon: A line graph connecting the mid-points of each class in a data set, plotted as a height corresponding to the frequency of the class.

fresh weight: The weight, including the water content, of a plant or plant part at the time of harvest.

friable callus: It is a crumbling or fragmenting callus which is easily dissected and readily dispersed into single cells or clumps of cells in a solution.

fruit fly (*Drosophila melanogaster*): It is a model organism for research for almost a century, and is considered an attractive system for a number of reasons, including its easy-to-manipulate genetic system, relatively low cost, and biological complexity comparable to that of a mammal. Many organ systems in mammals have well-conserved homologs in *Drosophila*, and *Drosophila* research has already led the way in providing new insights into forms of cancer, neurodegenerative diseases, behaviour, immunity, aging, multigenic inheritance, and development.

functional genomics: It refers to the use of genomic information to delineate protein structure, function, pathways and networks. Function may be determined by "knocking out" or "knocking in" expressed genes in model organisms such as worm, fruitfly, yeast or mouse.

fungicide: A chemical that kills fungi. Fungicides can either be contact or systemic. A contact fungicide kills fungi by direct contact; a systemic fungicide has to be absorbed by the affected organism.

furfural (furfuraldehyde): Used industrially as a solvent and raw material for synthetic resin. It is called Furfurol; 2-Furaldehyde; 2-Furancarboxaldehyde; Furan-2-carboxaldehyde. Its molecular formula is $\mathbf{C_5H_4O_2}$.

fusion (hybrid) gene: A hybrid gene is created by joining portions of two different genes (to produce a new protein), or by joining a gene to a different promoter fusion protein A polypeptide made from a recombinant gene that contains portions of two or more different genes. The different genes are joined so that their coding sequences are in the same reading frame: the genetic apparatus reads the gene fusion as a hybrid protein or chimeric protein.

fusion biopharmaceuticals: Biopharmaceutical proteins formed as a result of fusion proteins. An immunotoxin is a fusion biopharmaceutical and a man-made protein that consists of a targeting portion linked to a toxin. When the protein binds to that cell, it is taken in through endocytosis, and the toxin kills the cell. These are used for the treatment of some kinds of cancer and a few viral infections.

fusion protein: It refers to the protein resulting from genetic joining and expression of 2 different genes.

G

G cap: A *5' cap* (also termed an RNA cap, an RNA 7-methylguanosine cap or an RNA m^7G cap) is a modified guanine nucleotide that has been added to the "front" or 5' end of a eukaryotic messenger RNA shortly after the start of transcription. The 5' cap consists of a terminal 7-methylguanosine residue which is linked through a 5'-5'-triphosphate bond to the first transcribed nucleotide. Its presence is critical for recognition by the ribosome and protection from RNases. Cap addition is coupled to transcription, and occurs co-transcriptionally, such that each influences the other. Shortly after the start of transcription, the 5' end of the mRNA being synthesized is bound by a cap-synthesizing complex associated with RNA polymerase. This enzymatic complex catalyzes the chemical reactions that are required for mRNA capping. Synthesis proceeds as a multi-step biochemical reaction.

gametic equilibrium (linkage equilibrium): The condition in which the frequency of the gametes, formed by the association of alleles at different loci, is equal to the product of the frequencies of the alleles that constitute them.

gametoclone: A plant regenerated from a tissue culture originating from gametic tissue.

gametogenesis: The process of thc formation of gametes.

gametophyte: That phase of the plant life cycle, which bears the sex organs producing gametes with haploid number of chromosomes. In some algae and all bryophytes, gametophyte is the main plant body.

gap dynamics: It refers to the ecological effect of a canopy gap opened when a tree dies or is otherwise removed from a section of forest.

gap penalties: It refers to the penalty applied to a similarity score for the introduction of an insertion or deletion gap, the extension of a gap, or both. Gap penalties are usually subtracted from a cumulative score being determined for the comparison of two or more sequences via an optimization algorithm that attempts to maximize that score.

gapped DNA: A duplex DNA molecule with one or more internal single-stranded regions.

gaps (affine gaps): It refers to any maximal, consecutive run of spaces in a single string of a given alignment. Gaps help create alignments that better conform to underlying biological models and more closely fit patterns that one expects to find in meaningful alignment. The idea is to take in account the number of continuous gaps and not only the number of spaces when calculating an alignment. Affine gaps contain a component for gap insertion and a component for gap extension, where the extension penalty is usually much lower than the insertion penalty. This mimics biological reality as multiple gaps would imply multiple mutations, but a single mutation can lead to a long gap quite easily.

GC island: A segment of DNA that is rich in G=C base pairs and often precedes a transcribed gene in the genomes of vertebrate organisms.

GDP: Guanosine diphosphate.

gel A: lyophilic colloid that has coagulated to a rigid or jelly-like solid. It is used for the electrophoretic separation of nucleic acids or proteins, and for encapsulation.

gel electrophoresis: A method to analyze the size of DNA (or RNA) fragments. In the presence of an electric field, larger fragments of DNA move through the gel slower than the smaller ones. If a sample contains fragments at four different discrete sizes, those four size classes will, when subjected to electrophoresis, migrate in groups, producing four migrating "brands". Usually, these are visualized by soaking the gel in a dye (ethidium bromide), which makes the DNA fluoresce under UV light.

gelatin: It is a translucent, colourless, odourless, brittle, nearly tasteless solid substance derived from the collagen inside animals' skin and bones. It is commonly used as a gelling agent in food, pharmaceuticals, photography, and cosmetic manufacturing. Substances containing gelatin or functioning in a similar way are called *gelatinous*. Gelatin is an irreversibly hydrolyzed form of collagen, and is classified as a foodstuff. Gelatin is a protein produced by partial hydrolysis of collagen extracted from the bones, connective tissues, organs and some intestines of animals such as domesticated cattle, pigs, and horses. The natural molecular bonds between the individual collagen strands are broken down into a form that rearranges more easily. It melts to a liquid when heated and solidifies when cooled again. Together with water, it forms a semi-solid colloid gel.

GenBank: Data bank of genetic sequences operated by a division of the National Institutes of Health.

gene addition: The addition of a functional copy of a gene to the genome of an organism.

gene amplification: The process by which a cell specifically increases the copy number of a particular gene to a greater extent than it increases the copy number of genes composing the remainder of the genome (all the genes which make up the genetic machinery of an organism). It is therefore distinguished from duplication, which is a precise doubling of the genome preparatory to cell division, and endoreduplication, which leads to endopolyploidy. Gene amplification results from repeated replication of deoxyribonucleic acid (DNA) in a limited portion of the genome, in the absence of or to a much greater extent than replication of DNA composing the remainder of the genome. Thus is formed a cell in which the genes composing a limited portion of the genome are present in relatively high copy number, while the genes composing the remainder of the genome are present in approximately normal copy number. (*See also* Deoxyribonucleic acid (DNA)). Since gene amplification increases the copy number of a specific region of the genome without altering the copy number of genes composing the remainder of the genome, it would appear to offer an alternative method for developmental control of gene expression. By increasing the number of copies of a particular gene, the number of gene copies available for transcription could thereby be increased. In a number of instances of gene amplification, the amplification phenomenon appears to be developmentally regulated, and the amplified copies of the gene are subsequently lost from the cell. Studies on cells in culture have demonstrated "amplification" of genes involved in resistance to specific drugs.

gene bank: It is a means of preserving genetic material, be it plant or animal. In plants, this could be done by freezing cuts from the plant, or stocking the seeds. GenBank is the NIH genetic sequence database, an annotated collection of all publicly available DNA sequences. A new release is made every two months. GenBank is part of the International Nucleotide Sequence Database Collaboration, which is comprised of the DNA DataBank of Japan (DDBJ), the European Molecular Biology Laboratory (EMBL), and GenBank at the National Center for Biotechnology Information.

gene chips (gene arrays): It refers to the covalent attachment of oligonucleotides or cDNA directly onto a small glass or silicon chip in organized arrays. Over 50,000 different DNA fragments can be presented on a single chip providing a high throughput parallel method of probing gene expression, genotype or gene function.

gene cloning: The process of synthesizing multiple copies of a particular DNA sequence using a bacteria cell or another organism as a host. The gene of interest is inserted into a self-replicating DNA molecule (DNA vector, often a plasmid), and the resulting recombinant DNA molecule is amplified in an appropriate host cell.

gene conversion: A process, often associated with recombination, during which one allele is replicated at the expense of another, leading to non-Mendelian segregation ratios.

gene expression: It is a process by which information from a gene is used in the synthesis of a functional gene product. These products are often proteins, but in non-protein coding genes such as rRNA genes or tRNA genes, the product is a functional RNA. The process of gene expression is used by all known life—eukaryotes (including multicellular organisms), prokaryotes (bacteria and archaea) and viruses—to generate the macromolecular machinery for life. Several steps in the gene expression process may be modulated, including the transcription, RNA splicing, translation, and post-translational modification of a protein. Gene regulation gives the cell control over structure and function, and is the basis for cellular differentiation, morphogenesis and the versatility and adaptability of any organism. Gene regulation may also serve as a substrate for evolutionary change, since control of the timing, location, and amount of gene expression can have a profound effect on the functions (actions) of the gene in a cell or in a multicellular organism.

gene families: Subsets of genes containing homologous sequences which usually correlate with a common function.

gene flow: It is the transfer of alleles of genes from one population to another. Migration into or out of a population may be responsible for a marked change in allele frequencies (the proportion of members carrying a particular variant of a gene). Immigration may also result in the addition of new genetic variants to the established gene pool of a particular species or population.

gene frequency (allele frequency): It is the proportion of all copies of a gene that is made up of a particular gene variant (allele). In other words, it is the number of copies of a particular allele divided by the number of copies of all alleles at the genetic place (locus) in a population. It can be expressed, for example, as a percentage.

gene imprinting: Genetic phenomenon whereby a small number of genes function differently depending on whether they were inherited from the father or the mother. If two copies of an imprinted gene are inherited from one parent and none from the other, a genetic abnormality results, whereas no abnormality occurs if, as is normal, a copy is inherited from both parents. Gene imprinting is known to play a part in a number of genetic disorders and childhood diseases, for example, the Prader-Willi syndrome (characterized by mild mental retardation and compulsive eating). Imprinted genes play a part in the development of embryo and placenta.

gene index: It refers to the listing of the number, type, label and sequence of all the genes identified within the genome of a given organism. Gene indices are usually created by assembling overlapping EST sequences into clusters, and then determining if each cluster corresponds to a unique gene. Methods by which a cluster can be identified as representing a unique gene include identification of long open reading frames (ORFs), comparison to genomic sequence, and detection of SNPs or other features in the cluster that are known to exist in the gene.

gene insertion: It is the addition of one or more nucleotide base pairs into a DNA sequence. This can often happen in microsatellite regions due to the DNA polymerase slipping. Insertions can be anywhere in size from one base pair incorrectly inserted into a DNA sequence to a section of one chromosome inserted into another. On a chromosome level, an *insertion* refers to the insertion of a larger sequence into a chromosome. This can happen due to unequal crossover during meiosis. N-region addition is the addition of non-coded nucleotides during recombination by terminal deoxynucleotidyl transferase. P-nucleotide insertion is the insertion of palindromic sequences encoded by the ends of the recombining gene segments.

gene interaction: The elegance of Mendel's experiments was partly due to the complete consistency between his observations and the hypothesis he developed. Mendel was lucky that in his studies the characteristics he chose were governed by a single gene. However, after Mendel's work was rediscovered, it became clear that simple Mendelian model is not sufficient to predict experimental observations in all situations. Two or more genes govern the expression of many characters in almost all organisms. These genes affect the development of the concerned characters in various ways. The phenomena of two or more genes governing the development of a single character in such a way that they affect the expressions of each other in various ways is known as gene interaction. Gene interaction may involve two or more genes. But all the examples described below relate to interaction between two genes. Therefore, these cases are also referred to as the modification of dihybrid ratio.

gene library (genomic library): It is a population of host bacteria, each of which carries a DNA molecule that was inserted into a cloning vector, such that the collection of cloned DNA molecules represents the entire genome of the source organism. This term also represents the collection of all of the vector molecules, each carrying a piece of the chromosomal DNA of the organism, prior to the insertion of these molecules into the host cells. The DNA molecules of an organism of interest are isolated. The DNA molecules are then partially digested by an endonuclease restriction enzyme. Sometimes, the DNA molecules are digested for different lengths of time in order to ensure that all the DNA has been digested to manageable sizes. The digested DNA molecules are separated by size using agarose electrophoresis, and a suitable range of lengths of DNA pieces are isolated and ligated into vectors. The vectors can then be taken up by suitable hosts. The hosts are kept in liquid media and can be frozen at –80°C for a long period of time. Usually, the hosts are bacteria that do not contain any plasmids, so as to be sensitive to antibiotics. The process of subdividing genomic DNA into clonable elements and inserting them into hosts is called creating a library, a clone bank or a gene bank. A complete library of host cells will contain all of the genomic DNA of the source organism.

gene library: A collection of cloned DNA fragments created by restriction endonuclease digestion that represents part or all of an organism's genome.

gene linkage: It occurs when particular genetic loci or alleles for genes are inherited jointly. Genetic loci on the same chromosome are physically close to one another and tend to stay together during meiosis, and are thus genetically *linked.* This is called autosomal linkage. Alleles for genes on different chromosomes are usually not linked, due to independent assortment of chromosomes during meiosis. Because there is some crossing over of DNA when the chromosomes segregate, alleles on the same chromosome can be separated and go to different daughter cells. There is a greater probability of this happening if the alleles are far apart on the chromosome, as it is more likely that a crossover will occur between them. The relative distance between two genes can be calculated using the offspring of an organism showing two linked genetic traits, and finding the percentage of the offspring where the two traits do not run together. The higher the percentage of descendants that does not show both traits, the farther apart on the chromosome the two genes are. Among the individuals of an experimental population or species, some phenotypes or traits occur randomly with respect to one another in a manner known as independent assortment.

gene modification: Genetic engineering is different from traditional breeding, where the organism's genes are manipulated indirectly. Genetic engineering uses the techniques of molecular cloning and transformation to alter the structure and characteristics of genes directly. Genetic engineering techniques have found success in numerous applications. Some examples are in improving crop technology, the manufacture of synthetic human insulin through the use of modified bacteria, the manufacture of erythropoietin in hamster ovary cells, and the production of new types of experimental mice such as the oncomouse (cancer mouse) for research. The term "genetic engineering" was coined in Jack Williamson's science fiction novel *Dragon's Island*, published in 1951, two years before James Watson and Francis Crick showed that DNA could be the medium of transmission of genetic information.

gene pool: A large gene pool indicates extensive genetic diversity, which is associated with robust populations that can survive bouts of intense selection. Meanwhile, low genetic diversity can cause reduced biological fitness and an increased chance of extinction. When all individuals in a population are identical with regard to a particular phenotypic trait, they are known as monomorphic. When the individuals show several variants of a particular trait, they are said to be polymorphic.

gene probe: Probe is any device or agent, such as a radioactively tagged isotope or a molecular deoxyribonucleic acid fragment probe, inserted into a medium to obtain information about a structure or substance. It is a single-stranded DNA or RNA fragment used in genetic engineering to search for a particular gene or other DNA sequence. The probe has a base sequence complementary to the target sequence and will thus attach to it by base pairing. By labelling the probe, it can be identified after subsequent separation and purification.

gene product: The product, either RNA or protein, that results from expression of a gene. The amount of gene product reflects the activity of the gene.

gene recombination: It is a process by which a molecule of nucleic acid (usually DNA, but can also be RNA) is broken and then joined to a different DNA molecule. Recombination can occur between similar molecules of DNA, as in homologous recombination, or dissimilar molecules of DNA, as in non-homologous end joining. Recombination is a common method of DNA repair in both bacteria and eukaryotes. In eukaryotes, recombination occurs in meiosis as a way of facilitating chromosomal crossover. The crossover process leads to offspring having different combinations of genes from their parents, and can occasionally produce new chimeric alleles.

gene replacement: The incorporation of a transgene into a chromosome at its normal location by homologous recombination, thus replacing the copy of the gene originally present at the locus.

gene sequencing: The term refers to sequencing methods for determining the order of the nucleotide bases—adenine, guanine, cytosine, and thymine—in a molecule of DNA. Knowledge of DNA sequences has become indispensable for basic biological research, other research branches utilizing DNA sequencing, and in numerous applied fields such as diagnostic, biotechnology or forensic biology. The advent of DNA sequencing has significantly accelerated biological research and discovery. The rapid speed of sequencing attained with modern

DNA sequencing technology has been instrumental in the sequencing of human genome, in the Human Genome Project. Related projects, often by scientific collaboration across the continents, have generated the complete DNA sequences of many animals, plants, and microbial genomes.

gene shears: Technique in genetic engineering which may have practical applications in the future. The gene shears are pieces of messenger RNA that can bind to other pieces of messenger RNA, recognizing specific sequences, and cut them at that point. If a piece of DNA which codes for the shears can be inserted in the chromosomes of a plant or animal cell, that cell will then destroy all messenger RNA of a particular type. Genetic shears may be used to protect plants against viruses which infect them and cause disease. They might also be useful against AIDS.

gene splicing: A stage in the processing of mRNA, occurring only in eukaryote cells, in which intervening sequences (introns) are removed from the primary RNA transcript (hnRNA), and the coding exons are joined together to form the mature mRNAmolecule. Sliceosomes are the molecules involved in the process of splicing.

gene therapy: Genes, which are carried on chromosomes, are the basic physical and functional units of heredity. Genes are specific sequences of bases that encode instructions on how to make proteins. Although genes get a lot of attention, it is the proteins that perform most life functions and even make up the majority of cellular structures. When genes are altered so that the encoded proteins are unable to carry out their normal functions, genetic disorders can result. Gene therapy is a tcchniquc for correcting defective genes responsible for disease development. Researchers may use one of the several approaches for correcting faulty genes. A normal gene may be inserted into a nonspecific location within the genome to replace a nonfunctional gene. This approach is most common. An abnormal gene could be swapped for a normal gene through homologous recombination. The abnormal gene could be repaired through selective reverse mutation, which returns the gene to its normal function.

gene therapy: The use of genetic material for therapeutic purposes. The therapeutic gene is typically delivered using recombinant virus or liposome-based delivery systems.

gene tracking: Following a gene through a family by observing the inheritance of a marker which is being inherited alongside the gene.

gene translocation: The movement of a gene fragment from one chromosomal location to another, which often alters or abolishes the expression.

gene: It is the basic unit of heredity in a living organism. All living things depend on genes. Genes hold the information to build and maintain their cells and pass genetic traits to the offspring. A modern working definition of a gene is "a locatable region of genomic sequence, corresponding to a unit of inheritance, which is associated with regulatory regions, transcribed regions, and/or other functional sequence regions". Incorrect colloquial usage of the term gene may actually refer to an allele: a gene is the basic instruction, a sequence of DNA, while an allele is one variant of that instruction. The notion of a gene is evolving with the science of genetics, which began when Gregor Mendel noticed that biological variations are inherited from parent organisms as specific, discrete traits. The biological entity responsible for defining traits was termed a gene, but the biological basis for inheritance remained unknown until DNA was identified as a genetic material in the 1940s. All organisms have many genes corresponding to many different biological traits, some of which are immediately visible, such as eye colour or number of limbs, and some of which are not, such as blood type or increased risk for specific diseases, or the thousands of basic biochemical processes that comprise life. In cells, a gene is a portion of DNA that contains both "coding" sequences that determine what the gene does, and "non-coding" sequences that determine when the gene is active (expressed). When a gene is active, the coding and non-coding sequences are copied in a process called transcription, producing an RNA copy of the gene's information. This piece of RNA can then direct the synthesis of proteins via the genetic code. In other cases, the RNA is used directly, for example, as part of the ribosome.

generation time (cell generation time): The time required for a cell to complete one full growth cycle. If every cell in the population is capable of forming two daughter cells, has the same average generation time, and is not lost through lysis, the doubling time of the cell number in a population will equal the generation time.

genetic (DNA) fingerprinting: It is a technique employed by forensic scientists to assist in the identification of individuals on the basis of their respective DNA profiles. DNA profiles are encrypted sets of numbers that reflect a person's DNA makeup, which can also be used as the person's identifier. DNA profiling should not be confused with full genome sequencing. This technique was first reported in 1984 by Sir Alec Jeffreys at the University of Leicester in England.

genetic algorithm (GA): It is a search heuristic that mimics the process of natural evolution. This heuristic is routinely used to generate useful solutions to optimization and search problems. Genetic algorithms belong to the larger class of evolutionary algorithms (EA), which generate solutions to optimization problems using techniques inspired by natural evolution, such as inheritance, mutation, selection, and crossover.

genetic assimilation: It is a process by which the effect of an environmental condition, such as exposure to a teratogen, is used in conjunction with artificial selection or natural selection to create a strain of organisms with similar changes in phenotype that are encoded genetically. Despite superficial appearances, this does not require the inheritance of acquired characters, although epigenetic inheritance could potentially influence the result. Genetic assimilation is merely a method of overcoming the barrier to selection imposed by genetic canalization of developmental pathways.

genetic code: It is the set of rules by which information encoded in genetic material (DNA or RNA sequences) is translated into proteins (amino acid sequences) by living cells. A more precise term for the concept might be "genetic cipher". The code defines a mapping between tri-nucleotide sequences, called codons, and amino acids. A triplet codon in a nucleic acid sequence usually specifies a single amino acid (though in some cases the same codon triplet in different locations can code unambiguously for two different amino acids, the correct choice at each location being determined by context). Because the vast majority of genes are encoded with exactly the same code, this particular code is often referred to as the canonical or standard genetic code, or simply the genetic code, though in fact there are many variant codes. Thus, the canonical genetic code is not universal. For example, in humans, protein synthesis in mitochondria relies on a genetic code that varies from the canonical code. It is important to know that not all genetic information is stored using the genetic code. All organisms' DNA contain regulatory sequences, intergenic segments, and chromosomal structural areas that can contribute greatly to phenotype, but operate using distinct sets of rules that may or may not be as straightforward as the codon-to-amino acid paradigm that usually underlies the genetic code. *(See Appendix 2)*

genetic complementation: It refers to a relationship between two different strains of an organism, both having homozygous recessive mutations that produce the same phenotype.

genetic disease: It is an illness caused by abnormalities in genes or chromosomes. While some diseases, such as cancer, are due in part to a genetic disorder, they can also be caused by environmental factors. Most disorders are quite rare and affect one person in every several thousands or millions. Some types of recessive gene disorders confer an advantage in the heterozygous state in certain environments. (*See Appendix 3 for* ***Common genetic disorders***)

genetic distance: It refers to the genetic divergence between species or populations within a species. It considers a variety of parameters used to measure the genetic distance. Smaller genetic distances indicate a close genetic relationship, whereas large genetic distances indicate a more distant genetic relationship. Genetic distance can be used to compare the genetic similarity between different species, such as humans and chimpanzees.

genetic diversity: It is a level of biodiversity that refers to the total number of genetic characteristics in the genetic makeup of a species. It is distinguished from genetic variability, which describes the tendency of genetic characteristics to vary.

genetic drift (allelic drift): It is the change in the relative frequency in which a gene variant (allele) occurs in a population due to random sampling and chance. The alleles in the offspring are a random sample of those in the parents, and chance has a role in determining whether a given individual survives and reproduces.

genetic engineering: Changes in the genetic constitution of cells resulting from the introduction or elimination of specific genes through modern molecular biology techniques. This technology is based on the use of a vector for transferring useful genetic information from a donor organism into a cell or organism that does not possess it.

genetic equilibrium: In a population, genetic equilibrium is a condition in which the frequencies of allelic genes are maintained at the same values from generation to generation. It occurs when an allele within a gene pool is not changing in frequency.

genetic heterogeneity: The presence of apparently similar characters for which the genetic evidence indicates that different genes or different genetic mechanisms are involved in different pedigrees. In clinical settings, genetic heterogeneity refers to the presence of a variety of genetic defects which cause the same disease, often due to mutations at different loci on the same gene—a finding common to many human diseases including alzheimer disease; cystic fibrosis; lipoprotein lipase deficiency, familial; and polycystic kidney diseases.

genetic immunization: To produce an immune reaction against a foreign protein usually requires purification of that protein, which is then injected into an animal. The isolation of enough pure protein is time-consuming and sometimes difficult. Here we report that such a response can also be elicited by introducing the gene encoding a protein directly into the skin of mice. This is achieved using a hand-held form of biolistic system, which can propel DNA-coated gold microprojectiles directly into cells in the living animal. Genetic immunization may be time- and labour-saving in producing antibodies and may offer a unique method for vaccination.

genetic information: The heritable biological information coded in the nucleotide sequences of DNA or RNA (certain viruses), such as in the chromosomes or plasmids.

genetic linkage: It is a phenomenon which occurs when particular genetic loci or alleles for genes are inherited jointly. Genetic loci on the same chromosome are physically close to one another and tend to stay together during meiosis, and are thus genetically *linked*. This is called autosomal linkage. Alleles for genes on different chromosomes are usually not linked, due to independent assortment of chromosomes during meiosis.

genetic mapping (genome mapping): It is the creation of a genetic map assigning DNA fragments to chromosomes. When a genome is first investigated, this map is nonexistent. The map improves with the scientific progress and is perfect when the genomic DNA sequencing of the species has been completed. During this process, and for the investigation of differences in strain, the fragments are identified by small tags. These may be genetic markers or the unique sequence-dependent pattern of DNA-cutting enzymes. The ordering is derived from genetic observations for these markers or in the second case from a computational integration of the fingerprinting data. The term "mapping" is used in two different but related contexts. Two different ways of mapping are distinguished. Genetic mapping uses classical genetic techniques to determine sequence features within a genome.

genetic marker: It is a gene or DNA sequence with a known location on a chromosome and associated with a particular gene or trait. It can be described as a variation, which may arise due to mutation or alteration in the genomic loci, that can be observed. A genetic marker may be a short DNA sequence, such as a sequence surrounding a single base-pair change (single nucleotide polymorphism, SNP), or a long one, like minisatellites.

genetic marker: It refers to any gene that can be readily recognized by its phenotypic effect, and which can be used as a marker for a cell, chromosome, or individual carrying that gene. Also, any detectable polymorphism used to identify a specific gene.

genetic polymorphism: A difference in DNA sequence among individuals, groups, or populations that gives rise to different forms such as the human blood groups.

genetic resources conservation: The conservation of species, populations, individuals or parts of individuals, by in situ or ex situ methods, to provide a diversity of genetic materials for present and future generations.

genetic selection: The process of selecting genes, cells, clones, etc., within the populations or between populations or species. Genetic selection usually results in differential success rates of different genotypes, reflecting many variables, including selection pressure and genetic variability in populations.

genetic transformation: It is the genetic alteration of a cell resulting from the uptake, genomic incorporation, and expression of foreign genetic material (DNA). Transformation was first demonstrated in 1928 by British bacteriologist Frederick Griffith. Griffith discovered that a harmless strain of *Streptococcus pneumoniae* could be made virulent after being exposed to heat-killed virulent strains.

genetic variation: Variation in alleles of genes occurring both within and among populations. Genetic variation is important because it provides the "raw material" for natural selection.

genetically modified organism (genetically engineered organism): It is an organism whose genetic material has been altered using genetic engineering techniques. These techniques, generally known as recombinant DNA technology, use DNA molecules from different sources, which are combined into one molecule to create a new set of genes. This DNA is then transferred into an organism, giving it modified or novel genes. Transgenic organisms, a subset of GMOs, are organisms which have inserted DNA that originated in a different species. Some GMOs contain no DNA from other species and are therefore not transgenic but cisgenic. Genetic modification involves the insertion or deletion of genes. When genes are inserted, they usually come from a different species, which is a form of horizontal gene transfer. In nature this can occur when exogenous DNA penetrates the cell membrane for any reason. To do this artificially may require attaching the genes to a virus or just physically inserting the extra DNA into the nucleus of the intended host with a very small syringe, or with very small particles fired from a gene gun. However, other methods exploit natural forms of gene transfer, such as the ability of *Agrobacterium* to transfer genetic material to plants, or the ability of lentiviruses to transfer genes to animal cells.

genetics: The branch of science dealing with the study of heredity and variation.

genome: In modern molecular biology, the genome is the entirety of an organism's hereditary information. It is encoded either in DNA or, for many types of virus, in RNA. The genome includes both the genes and the non-coding sequences of the DNA. The term was adapted by Hans Winkler. The term genome can be applied specifically to mean that stored on a complete set of *nuclear DNA*, but can also be applied to that stored within organelles that contain their own DNA, as with the "mitochondrial genome" or the "chloroplast genome".

genomic DNA (sequence): It refers to the DNA sequence typically obtained from mammalian or other higher-order species, which includes both intron and exon sequence (coding sequence), as well as non-coding regulatory sequences such as promoter, and enhancer sequences.

genomic DNA library (genomic library): It is a population of host bacteria, each of which carries a DNA molecule that was inserted into a cloning vector, such that the collection of cloned DNA molecules represents the entire genome of the source organism. This term also represents the collection of all of the vector molecules, each carrying a piece of chromosomal DNA of the organism, prior to the insertion of these molecules into the host cells.

genomics: It refers to the analysis of the entire genome of a chosen organism.

genotype: It is the genetic constitution of a cell, an organism, or an individual (i.e., the specific allele makeup of the individual) usually with reference to a specific character under consideration.

geometric mean: A measure of central tendency used to measure the average rate of change or growth for some quality, computed by taking the nth root of the product of 'n' values representing change.

germ cell gene therapy: The repair or replacement of a defective gene within the gamete-forming tissues, resulting in a heritable change in an organism's genetic constitution.

germ cells: These are the cells that give rise to the gametes of organisms that reproduce sexually. In many animals, the germ cells originate near the gut and migrate to the developing gonads. There, they undergo cell division of two types, mitosis and meiosis, followed by cellular differentiation into mature gametes, either eggs or sperm. Unlike animals, plants do not have a germ line set aside in early development. Instead, germ cells can come from somatic cells in the adult floral meristem.

germ line cells: Cells that produce gametes.

germ line gene therapy: The delivery of a gene or genes to a fertilized egg or an early embryonic cell. The transferred genes are present in all or some of the nuclei of the cells of the mature individual, including possibly the reproductive cells, and alters the phenotype of the individual that develops.

germ line: It is the line of germ cells that has genetic material that may be passed to a child. For example, sex cells such as the sperm or the egg, are part of the germ line. So are the cells that produce sex cells, called gametocytes, the cells that produce those, called gametogonia, and all the way back to the zygote, the cell from which the individual developed.

germicide: Any chemical agent used to control or kill any pathogenic and non-pathogenic microorganisms.

germplasm: It is a collection of genetic resources for an organism. For plants, the germplasm may be stored as a seed collection or, for trees, in a nursery.

gibberellins(GAs): These are plant hormones that regulate the growth and influence various developmental processes, including stem elongation, germination, dormancy, flowering, sex expression, enzyme induction, and leaf and fruit senescence.

gibbons: It refers to small apes that use their arms to swing from branch to branch.

glaucus: A surface with a waxy, white coating. In most cases, this waxy covering can be rubbed off.

global warming: It refers to the rising of the earth's average global temperature because of greenhouse gases accumulating in the atmosphere.

globulin: It is one of the two types of serum proteins, the other being albumin. This generic term encompasses a heterogeneous group of proteins with typical high molecular weight, and both solubility and electrophoretic migration rates lower than for albumin. The normal concentration in blood is 2 to 3.5 g/dl. It is sometimes used synonymously with globular protein. However, albumin is also a globular protein, but not a globulin. All other serum globular proteins are globulins.

glucose invertase: Enzyme that catalyses the interconversion of two sugars, glucose and fructose. As fructose is chemically more stable than glucose, a mixture of glucose and fructose with the enzyme will end up almost entirely as fructose.

gluten: It is a composite of the proteins *gliadin* and *glutenin*. These exist, conjoined with starch, in the endosperms of some grass-related grains, notably wheat, rye, and barley. Gliadin and glutenin comprise about 80% of the protein contained in wheat seed. Being insoluble in water, they can be purified by washing away the associated starch. Worldwide, gluten is an important source of nutritional protein, both in foods prepared directly from sources containing it, and as an additive to foods otherwise low in protein.

glycolysis: It is the metabolic pathway that converts glucose, $C_6H_{12}O_6$, into pyruvate, $C_3H_3O_3^-$. The free energy released in this process is used to form the high energy compounds, ATP (adenosine triphosphate) and NADH (reduced nicotinamide adenine dinucleotide). Glycolysis is a definite sequence of ten reactions involving ten intermediate compounds.

glycosylation: It is the enzymatic process that links saccharides to produce glycans, attached to proteins, lipids, or other organic molecules. This enzymatic process produces one of the fundamental biopolymers found in cells (along with DNA, RNA, and proteins).

gobar gas (biogas): It typically refers to a gas produced by the biological breakdown of organic matter in the absence of oxygen. Biogas originates from biogenic material and is a type of biofuel.

Golgi apparatus: An assembly of vesicles and folded membranes within the cytoplasm of plant and animal cells. It was named after its discoverer, the Italian cytologist Camillo Golgi. It stores and transports secretory products and plays a role in the formation of cell wall.

goodness of fit test: A statistical test for determining whether there is a significant difference between an observed frequency distribution and a theoretical frequency distribution such as binomial, poisson, and normal.

G-proteins: These, also known as seven-transmembrane domain receptors, 7TM receptors, heptahelical receptors, serpentine receptors, and G protein-linked receptors (GPLR), comprise a large protein family of transmembrane receptors that sense molecules outside the cell and activate inside signal transduction pathways and, ultimately, cellular responses. G protein-coupled receptors are found only in eukaryotes, including yeast, choanoflagellates, and animals. The ligands that bind and activate these receptors include light-sensitive compounds, odours, pheromones, hormones, and neurotransmitters, and vary in size from small molecules to peptides to large proteins. G protein-coupled receptors are involved in many diseases, and are also the target of around half of all modern medicinal drugs.

graft inoculation test: A test based on the use of a suspected viral carrier which is grafted to an indicator plant. If symptoms appear in the indicator plant, the viral assay is positive.

grafting: It is a method of asexual plant propagation widely used in agriculture and horticulture where the tissues of one plant are encouraged to fuse with those of another. It is most commonly used for the propagation of trees and shrubs grown commercially. A physical weak point often still occurs at the graft, because the structural tissue of the two distinct plants, such as wood, may not fuse.

Gram staining (or Gram's method): It is an empirical method of differentiating bacterial species into two large groups (Gram-positive and Gram-negative) based on the chemical and physical properties of their cell walls. While Gram staining is a valuable diagnostic tool in both clinical and research settings, not all bacteria can be definitively classified by this technique, thus forming *Gram variable* and *Gram indeterminant* groups as well. The method is named after its inventor, the Danish scientist Hans Christian Gram (1853–1938), who developed the technique in 1884 to differentiate two types of bacteria with similar clinical symptoms: *Streptococcus pneumoniae* (also known as the *pneumococcus*) and *Klebsiella pneumoniae* bacteria. The word Gram is always spelled with a capital, referring to the name of the inventor of the Gram staining.

Gram-negative bacteria: These are the bacteria that do not retain crystal violet dye in the Gram staining protocol. In a Gram stain test, a counterstain (commonly safranin) is added after the crystal violet, colouring all Gram-negative bacteria with a red or pink colour. The test itself is useful in classifying two distinct types of bacteria based on the structural differences in their cell walls.

Gram-positive bacteria: These are the bacteria that are stained dark blue or violet by Gram staining. This is in contrast to Gram-negative bacteria, which cannot retain the crystal violet stain, and instead take up the counterstain (safranin or fuchsin) and appear red or pink.

graph: A graph is a visual form of representation of statistical data.

gratuitous inducer: It is a molecule which is structurally similar to another molecule that induces transcription for a specific product, and which can also initiate transcription for that product when the official inducer is absent.

green revolution: It refers to the transformation of agriculture that began in 1945, largely due to the life work of Norman Borlaug. One significant factor in this revolution was the Mexican government's request to establish an agricultural research station to develop more varieties of wheat that could be used to feed the rapidly growing population of the country. In 1943, Mexico imported half its wheat, but by 1956, the Green Revolution had made Mexico self-sufficient; by 1964, Mexico exported half a million tonnes of wheat.

greenhouse effect: It refers to the gradual warming of a planet by an atmosphere's conversion of incoming solar radiation into heat.

greenwashing: It refers to the political or business practice of masking an environmentally destructive activity by promoting it as environmentally beneficial.

group selection: It refers to the controversial idea that instead of responding only to challenges that presently face an organism, natural selection encourages improvements that will eventually benefit a group. This would lend evolution a forward-looking or "teleological" capacity most Darwinians reject.

growth cabinet: A cupboard used for incubating tubes or culture vessels under controlled environmental conditions. The degree of control over temperature, light and humidity is a function of the quality of the cabinet.

growth curve: It is an empirical model of the evolution of a quantity over time. Growth curves are widely used in biology for quantities such as population size, body height or biomass. Values for the measured property can be plotted on a graph as a function of time. Growth curves are employed in many disciplines besides biology, particularly in statistics, which has an extensive literature on "growth curves". Generation of a growth curve can be useful in evaluating the growth characteristics of a cell line during plant and animal cell cultures. From a growth curve, the lag time, population doubling time, and saturation density can be determined.

growth factor: It is a naturally occurring substance capable of stimulating cellular growth, proliferation and cellular differentiation. Usually, it is a protein or a steroid hormone. Growth factors are important for regulating a variety of cellular processes. Growth factors typically act as signalling molecules between cells.

growth hormone (GH): It is a protein-based polypeptide hormone. It stimulates the growth and cell reproduction and regeneration in humans and other animals. It is a 191-amino acid, single-chain polypeptide hormone that is synthesized, stored, and secreted by the somatotroph cells within the lateral wings of the anterior pituitary gland. Somatotropin refers to the growth hormone produced natively and naturally in animals, whereas the term somatropin refers to growth hormone produced by recombinant DNA technology, and is abbreviated "rhGH" in humans.

growth inhibitor: Any substance inhibiting the growth of an organism. The inhibitory effect can range from mild inhibition (growth retardation) to severe inhibition or death (toxic reaction). Two plant growth regulators that may act as inhibitors are ethylene and abscisic acid. The concentration of the inhibitor, the length of exposure to it, and the relative susceptibility of the organisms exposed to the inhibitor are important factors which determine the extent of the inhibitory effect.

growth rate: Year-over-year change in the height and weight of the plant, expressed as a percentage is referred to as growth rate.

growth regulator (plant hormones or phytohormones): These are the chemicals that regulate plant growth. In UK, these are termed 'plant growth substances'. Plant hormones are signal molecules produced within the plant, and occur in extremely low concentrations. Hormones regulate cellular processes in targeted cells locally and when moved to other locations, in other locations of the plant. Hormones also determine the formation of flowers, stems, leaves, the shedding of leaves, and the development and ripening of fruit. Plants, unlike animals, lack glands that produce and secrete hormones, instead each cell is capable of producing hormones. Plant hormones shape the plant, affecting seed growth, time of flowering, the sex of flowers, senescence of leaves and fruits. They affect which tissues grow upward and which grow downward, leaf formation and stem growth, fruit development and ripening, plant longevity, and even plant death. Hormones are vital to plant growth and lacking them, plants would be mostly a mass of undifferentiated cells.

growth retardant: A chemical that selectively interferes with normal hormonal promotion of growth and other physiological processes, but without appreciable toxic effects.

guanine: A purine derivative that is one of the major component bases of nucleotides and the nucleic acids, DNA and RNA.

guanosine: A nucleoside consisting of one guanine molecule linked to a D-ribose sugar molecule. The derived nucleotides, guanosine mono-, di-, and triphosphate (GMP, GDP, and GTP, respectively), are important in various metabolic reactions.

guide RNA (gRNA): These are the RNAs that guide the insertion or deletion of uridine residues into mitochondrial mRNAs in kinetoplastid protists in a process known as RNA editing.

guide sequence: An RNA molecule (or a part of it) which hybridizes with eukaryotic mRNA and aids in the splicing of intron sequences. Guide sequences may be either external (EGS) or internal (IGS) to the RNA being processed and may hybridize with either intron or exon sequences close to the splice junction.

guild: It refers to a group of species that fill similar ecological niches.

gyrase (DNA gyrase): It is an enzyme that unwinds double-stranded DNA. This causes supercoiling of the DNA. Many antibiotics work by attacking bacterial DNA gyrase.

H

habituation: It is the psychological process in humans and animals in which there is a decrease in psychological response and behavioural response to a stimulus after repeated exposure to that stimulus over a duration of time. Habituation is very similar to acclimation in that repetition of certain behaviours that are rewarding to a life form will likely be continued, or ingrained in a habitual manner.

haemoglobin: It is the iron-containing oxygen-transport metalloprotein in the red blood cells of vertebrates, and the tissues of some invertebrates. In mammals, the protein makes up about 97% of the red blood cell's dry content, and around 35% of the total content (including water). Haemoglobin transports oxygen from the lungs or gills to the rest of the body (i.e., the tissues) where it releases the oxygen for cell use. It also has a variety of other roles of gas transport and effect-modulation, which vary from species to species, and are quite diverse in some invertebrates. For example, haemoglobin transports CO_2 back from the tissues to the lungs. Haemoglobin has an oxygen binding capacity of between 1.36 and 1.37 ml O_2 per gram of haemoglobin, which increases the total blood oxygen capacity seventyfold. Haemoglobin is found in non-erythroid cells including the A9 dopaminergic neurons in the substantia nigra, macrophages, alveolar cells, and mesangial cells in the kidney. In these tissues, it has a non-oxygen carrying function as an antioxidant and a regulator of iron metabolism.

haemolymph: It is the blood analogue used by most arthropods and mollusks that have an open circulatory system. In these animals there is no distinction between blood and interstitial fluid. The liquid fills all of the interior (the hemocoel) of the body and surrounds all cells. It is distinguished from most animals and human blood by the absence of red blood cells.

hairpin: It refers to a double-helical region in a single DNA or RNA strand formed by the hydrogen-bonding between adjacent inverse complementary sequences to form a hairpin-shaped structure.

hairy root culture: It is a strategy used in plant tissue culture to produce good quantities of valuable secondary metabolites. An infection of plants with the soil bacterium *Agrobacterium rhizogenes* (Riker et al., 1930, Young et al., 2001) leads to a genetic transformation of plant cells. *Agrobacterium rhizogenes* causes the hairy root disease in plants. The neoplastic (cancerous) roots produced by *A. rhizogenes* infection are characterized by high growth rate, genetic stability and growth in hormone-free media. These genetically transformed root cultures can produce high levels of secondary metabolites comparable to that of intact plants.

hairy root disease: A disease of broad-leaved plants, where a proliferation of root-like tissue is formed from the stem. Hairy root disease is a tumorous state similar to crown-gall, and is induced by the bacterium *Agrobacterium rhizogenes,* containing an Ri plasmid. (*See Agrobacterium*, crown gall).

half-life: It refers to time required for half of a quantity of particle-emitting material to undergo radioactive decay. Because emission slows as the isotope decreases in radioactivity, taking half the life of its decay is more meaningful than trying to estimate the entire amount. Half-lives range from millionths of a second to millions of years.

halothane (or fluothane): It is an inhalational general anaesthetic. Its IUPAC name is 2-bromo-2-chloro-1,1,1-trifluoroethane. It is the only inhalational anaesthetic agent containing a bromine atom; there are several other halogenated anaesthesia agents which lack the bromine atom and do contain the fluorine and chlorine atoms present in halothane. It is colourless and pleasant-smelling, but unstable in light. It is packaged in dark-coloured bottles and contains 0.01% thymol as a stabilising agent.

hanging droplet technique (microdroplet array): It is technique used in the single cell cultures and is also used for germinating pollen grains in the laboratory. In this technique a drop of the nutrient medium is placed on the glass slide and the cells or pollen are inoculated on to the medium. The slide is then inverted over another slide with a cavity in such a way that the drop of the nutrient medium hangs over the cavity. Germination of pollen or cell can now be easily observed under microscope continuously.

haploid: A cell or organism containing only one representative from each of the pairs of homologous chromosomes found in the normal diploid cell. Having a single complete set of chromosomes, or referring to an individual or generation containing such a single set of chromosomes per cell (usually a gamete).

haplotype: It refers to a chunk of DNA that moves down the generations as a single unit.

haplotype: The term is a contraction of the term 'haploid genotype'. In genetics, a haplotype is a combination of alleles at multiple loci that are transmitted together on the same chromosome. Haplotype may refer to as few as one locus or an entire chromosome depending on the number of recombination events that have occurred between a given set of loci.

haptoglobin (Hp): It is a protein that in humans is encoded by the *HP* gene. In blood plasma, haptoglobin binds free haemoglobin (Hb) released from erythrocytes with high affinity and thereby inhibits its oxidative activity. The haptoglobin-hemoglobin complex will then be removed by the reticuloendothelial system (mostly the spleen). In clinical settings, the haptoglobin assay is used to screen for and monitor intravascular hemolytic anaemia. In intravascular hemolysis, free haemoglobin will be released into circulation and hence haptoglobulin will bind the Hb.

hard technology: It refers to costly, highly centralized, mechanical technology that uses a lot of energy, wastes resources, and pollutes the environment.

hardening (technique): It is a process by which an individual plant becomes tolerant to the effects of freezing during a period of weeks to months. It is a three-stage process. During the first stage, carbohydrates are translocated to the roots of the plant and cell membrane permeability increases. By the end of the first stage, the plant will be able to tolerate temperatures to –5 to –10°C. In the second stage, cell membranes are chemically altered, and this process will allow plants to survive temperatures of –20°C. In the third stage, which occurs in plants that survive at temperatures as low as –50°C, vitrification may occur.

Hardy-Weinberg equilibrium (principle): It states that both allele and genotype frequencies in a population remain constant—that is, they are in equilibrium—from generation to generation unless specific disturbing influences are introduced.

harmonic mean: A measure of central tendency used to measure an average rate like kilometre per hour, items manufactured per day etc., computing by taking reciprocal of the arithmetic mean of the reciprocal of the value of the variables.

heat pump: It is a machine or device that moves heat from one location (the 'source') to another using mechanical work.

helophyte (cryptophytes): It refers to a plant whose reproductive organs are underground or underwater that flourish in water or waterlogged soil.

helper plasmid: A plasmid that provides a function or functions to another plasmid in the same cell.

helper T cells (helper T lymphocytes or T_h cells: They are a sub-group of lymphocytes (a type of white blood cell or leukocyte) that play an important role in establishing and maximizing the capabilities of the immune system. These cells are unusual in that they have no cytotoxic or phagocytic activity; they cannot kill infected host (also known as somatic) cells or pathogens, and without other immune cells they would usually be considered useless against an infection. T_h cells are involved in activating and directing other immune cells, and are particularly important in the immune system. They are essential in determining B cell antibody class switching, in the activation and growth of cytotoxic T cells, and in maximizing bactericidal activity of phagocytes such as macrophages. It is this diversity in function and their role in influencing other cells that gives T helper cells their name.

helper virus: It is a virus used when producing copies of a helper dependent viral vector which does not have the ability to replicate on its own. The helper virus is used to co-infect cells alongside the viral vector and provides the necessary enzymes for replication of the genome of the viral vector.

hemicellulase: An enzyme that hydrolyzes pentosan (a type of hemicellulose), mainly used in the baking industry to improve the quality of dough; also used in food, beverage, animal feed, and pharmaceutical industries.

hemicelluloses: Any cellulose-like carbohydrate, but with differing chemical composition. Together with pectin and lignin, hemicelluloses form the cell wall matrix.

hemiphyte: It refers to an epiphyte (a plant that grows on another plant) that eventually extends its own roots into the ground.

hemizygous (haplozygous): The condition in which only one allele of a pair is present, as in sex linkage or as a result of deletion. Genes present only once in the genotype and not in pairs; as in haploids, in differential segments of sex chromosomes, or in diploids as a result of aneuploidy or loss of chromosome segments.

HEPA filter (high efficiency particulate air filter): It is generally used in inoculation chambers during microbial and tissue culture. It filters the air coming inside the chamber and makes the place sterile.

herbicide resistance: The ability of a plant to withstand herbicide. Herbicide resistance has been one of the early targets of plant genetic engineering.

herbicide: It is a substance used to kill unwanted plants. Selective herbicides kill specific targets while leaving the desired crop relatively unharmed. Some of these act by interfering with the growth of the weed and are often synthetic "imitations" of plant hormones.

heredity: It is the passing of traits to offspring (from its parent or ancestors). This is the process by which an offspring cell or organism acquires or becomes predisposed to the characteristics of its parent cell or organism. Through heredity, variations exhibited by individuals can accumulate and cause a species to evolve. The study of heredity in biology is called genetics, which includes the field of epigenetics.

heritability: It is the proportion of phenotypic variation in a population that is attributable to genetic variation among the individuals. Variation among individuals may be due to genetic and/or environmental factors. Heritability analyses estimate the relative contributions of differences in genetic and non-genetic factors to the total phenotypic variance in a population.

hermaphrodite: It is an animal or plant that has both male and female reproductive organs.

heteroalleles: Genes that have undergone mutation at different nucleotide positions and therefore result from different mutational events.

heterochromatin: It is a tightly packed form of DNA, which comes in two varieties: constitutive and facultative. Both play a role in the expression of genes: constitutive heterochromatin can affect the genes near them, and facultative heterochromatin is the result of genes that are silenced through a mechanism such as histone methylation or siRNA through RNAi. Constitutive heterochromatin is usually repetitive and forms structural functions such as centromeres or telomeres, in addition to acting as an attractor for other gene-expression or repression signals. Facultative heterochromatin is not repetitive, and although it shares the compact structure of constitutive heterochromatin, facultative heterochromatin can, under specific developmental or environmental signalling cues, lose its condensed structure and become a transcriptionally active gene.

heterodimer: It is a protein composed of 2 different chains or subunits.

heteroduplex: It is a double-stranded (duplex) molecule of nucleic acid originated through the genetic recombination of single complementary strands derived from different sources, such as from different homologous chromosomes or even from different organisms. One such example is the heteroduplex DNA strand formed in hybridization processes, usually for biochemistry-based phylogenetical analyses.

heterogametic sex: The gender which has two different sex chromosomes.

heterogeneous nuclear RNA (hnRNA): The original RNA transcripts found in eukaryotic nuclei before post-transcriptional modifications. A diverse assortment of RNA types found in the nucleus, including mRNA precursors (pre-mRNA) and other types of RNA.

heterokaryon: It is a cell that contains multiple, genetically different nuclei. This can occur naturally, such as in the mycelium of fungi during sexual reproduction, or artificially as formed by the experimental fusion of two genetically different cells.

heterologous expression: It means that a protein is experimentally put into a cell that does not normally express that protein.

heterologous interface: It is a subunit-subunit interface which occurs between different sequence regions of the protein subunit. All hetero-oligomeric interfaces are heterologous by definition, occurring between different subunits. However, homo-oligomeric interfaces are only rarely heterologous. In contrast, homo-oligomers tend to have isologous interfaces.

heterologous probe: A DNA probe that is derived from one species and is used to screen a similar DNA sequence from another species.

heterologous protein: A protein that differs from any protein normally found in the organism in question.

heterologous transplant: It refers to 'between species' or 'from one species to another'.

heteroplasmy: It is the presence of a mixture of more than one type of an organellar genome [mitochondrial DNA (mtDNA) or plastid DNA] within a cell or individual. Since every eukaryotic cell contains many hundreds of mitochondria with hundreds of copies of mtDNA, it is possible and indeed very frequent for mutations to affect only some of the copies, while the remaining ones are unaffected.

heteroploid: Having a chromosome number that is not a whole-number multiple of the haploid chromosome number for that species.

heteropyknosis: A state of variable density between the chromosomes of different cells or between individual chromosomes.

heterosis (hybrid vigour): It is a term used in genetics and selective breeding. It describes the increased strength of different characteristics in hybrids; the possibility to obtain a genetically superior individual by combining the virtues of its parents. Heterosis is the opposite of inbreeding depression, which occurs with increasing homozygosity.

heterozygote: An individual that has different alleles at the same locus in its two homologous chromosomes.

Hfr: It refers to high-frequency recombination strain of *Escherichia coli;* in such strains, the F episome is integrated into the bacterial chromosome.

hibernation: It refers to an energy-conserving lowering of metabolism, respiration, heart rate, and body temperature, while the body survives on accumulated fat supplies.

Hidden Markov model (HMM): It is a joint statistical model for an ordered sequence of variables. The result of stochastically perturbing the variables in a Markov chain (the original variables are thus "hidden"), where the Markov chain has discrete variables which select the "state" of the HMM at each step. The perturbed values can be continuous and are the "outputs" of the HMM. A Hidden Markov Model is equivalently a coupled mixture model where the joint distribution over states is a Markov chain. Hidden Markov models are valuable in bioinformatics because they allow a search or alignment algorithm to be trained using unaligned or unweighted input sequences, and because they allow position-dependent scoring parameters such as gap penalties, thus more accurately modeling the consequences of evolutionary events on sequence families.

high-throughput screening: It refers to the method by which very large numbers of compounds are screened against a putative drug target in either cell-free or whole-cell assays. Typically, these screenings are carried out in 96 well plates using automated, robotic station-based technologies or in higher-density array ("chip") formats.

histocompatibility: It is the property of having the same, or mostly the same, alleles of a set of genes called the major histocompatibility complex. These genes are expressed in most tissues as antigens, against which the immune system makes antibodies. The immune system at first makes antibodies against all sorts of antigens, including those it has never been exposed to, but stops making them against the antigens present in the body. If the body is exposed to foreign antigens, as by getting a tissue graft, it attacks the foreign material unless it is histocompatible.

histogram: Histogram is a set of vertical bars whose areas are proportional to the frequencies represented.

histone: Histones are strongly alkaline proteins found in eukaryotic cell nuclei, which package and order the DNA into structural units called nucleosomes. Without histones, the unwound DNA in chromosomes would be very long.

HIV: Human immunodeficiency virus. The retrovirus that causes AIDS in humans.

HLA (human leukocyte antigen): It is the name of the major histocompatibility complex (MHC) in humans. The superlocus contains a large number of genes related to immune system function in humans. This group of genes resides on chromosome 6, and encodes cell-surface antigen-presenting proteins and many other genes.

HLA complex: It is another name for the MHC in humans; refers to the "Human Leukocyte Antigen" complex located on chromosome 6.

holdfast: It refers to an anchor or "foot" used by an organism to stick to a surface.

hollow fibre: A tube of a porous material, with an internal diameter of a fraction of a millimetre. As a result, the ratio of surface area to internal volume is very large. Hollow fibres can be used as filters, and as they have a huge surface area, they take longer to clog up than normal filters. They are also used in the hollow fibre bioreactor, in which cells are placed inside the hollow, porous fibres, and the culture medium is circulated outside the reactor.

holoenzyme: The active compound formed by the combination of a coenzyme and an apoenzyme.

holometabolism: It is a term applied to insect groups to describe the specific kind of insect development, which includes four life stages—an embryo, a larva, a pupa and an imago or adult. Holometabolism is a monophyletic trait that all insects in the Endopterygota superorder have. This type of development gives the insects the unique advantage of being able to inhabit different ecological niches because of the morphological differences in the different stages of their life cycle.

homeobox (homeodomain): A homeobox is a DNA sequence found within the genes that are involved in the regulation of patterns of development (morphogenesis) in animals, fungi and plants. They were discovered independently in 1983 by Walter Jakob Gehring and his colleagues at the University of Basel, Switzerland, and Matthew P. Scott and Amy Weiner, who were then working with Thomas Kaufman at Indiana University in Bloomington.

homeobox: It refers to the highly conserved region in a homeotic gene composed of 180 bases (60 amino acids) that specifies a protein domain (the homeodomain) that serves as a master genetic regulatory element in cell differentiation during development in species as diverse as worms, fruitflies, and humans.

homeodomain: It refers to the 60 amino acid protein domain coded for by the homeobox region of a homeotic gene.

homeotic gene: It is a gene that controls the activity of other genes involved in the development of a body plan. Homeotic genes have been found in organisms ranging from plants to humans.

homeotic genes: These are the genes that determine which parts of the body form what body parts. One example is the *Hox* genes, which are important for segmentation, another example is the MADS-box-containing genes in the ABC model of flower development.

homeotic mutation: A mutation in homeobox genes resulting in homeotic transformation in animals. A mutation that causes a body part to develop in an inappropriate position in an organism, such as the mutation in *Drosophila* that causes legs to develop on the head in place of antennae.

homoalleles: Alternative forms of a gene that differ at the same muton site. Intragenic recombination between homoalleles is not possible.

homodimer: A protein made of paired identical polypeptides.

homogenization: It refers to gushing milk through a filter to spread out the fat globules into a mist.

homogenotization: It is a genetic technique used to replace one copy of a gene, or other DNA sequence within a genome, with an altered copy of that sequence. The DNA is first cloned and then altered in some way, e.g., a transposon is inserted into a gene.

homokaryon: A bi- or multinucleate cell having nuclei all of the same kind.

homologous chromosomes: These are the chromosomes in a biological cell that pair during cell division during the creation of gametes. Each chromosome pair contains genes for the same biological features, such as eye colour, at the same locations on the chromosome.

homologous recombination: It refers to the exchange of DNA fragments between two DNA molecules or chromatids of paired chromosomes at the site of identical nucleotide sequences.

homology modelling (comparative modelling): It refers to constructing an atomic-resolution model of the "target" protein from its amino acid sequence and an experimental three-dimensional structure of a related homologous protein (the "template"). Homology modelling relies on the identification of one or more known protein structures likely to resemble the structure of the query sequence, and on the production of an alignment that maps residues in the query sequence to residues in the template sequence. The sequence alignment and template structure are then used to produce a structural model of the target. Because protein structures are more conserved than DNA sequences, detectable levels of sequence similarity usually imply significant structural similarity.

homology: It refers to a phenomenon where, two or more biological species, systems or molecules share a common evolutionary ancestor. It also refers to two or more gene or protein sequences that share a significant degree of similarity, typically measured by the amount of identity (in the case of DNA), or conservative replacements (in the case of protein) that they register along their lengths. Sequence "homology" searches are typically performed with a query DNA or protein sequence to identify known genes or gene products that share significant similarity and hence might inform on the ancestry, heritage and possible function of the query gene.

homology: It refers to any similarity between characteristics of organisms, which is due to their shared ancestry. Anatomical structures that perform the same function in different biological species and evolved from the same structure in some ancestor species are homologous. In genetics, homology refers to a similarity of DNA sequences. The DNA sequences are usually similar, not identical. DNA sequences in the same cell can be similar, and DNA in different organisms can be similar. DNA codes with similar sequences are presumed to have a common ancestry. The concept of homology is contrasted with that of analogy, which refers to two structures that perform the same or similar function by a similar mechanism but evolved separately. Hence, similar structures may have evolved through different pathways, a process known as convergent evolution, or may be homologous.

homoplasmy: It is the presence of a mutation affecting all of the mitochondrial DNA (mtDNA) copies in a mammalian cell or chloroplast DNA in a plant cell. Since there are hundreds or even thousands of mtDNA copies in every eukaryotic cell, mutations may either be present in all copies (homoplasmy) or affect only a fraction of them (heteroplasmy).

homopolymer: A polymer resulting from the polymerization of a single monomer; a polymer consisting substantially of a single type of repeating unit.

homopolymeric tailing: In this example, a poly(G) tail is synthesized at each end of a blunt-ended DNA molecule. Tails comprising other nucleotides are synthesized by including the appropriate dNTP in the reaction mixture.

homozygote: A person who has two identical forms of a particular gene, one inherited from each parent.

hormone: It is a chemical released by one or more cells that affect cells in other parts of the organism. Only a small amount of hormone is required to alter cell metabolism. It is essentially a chemical messenger that transports a signal from one cell to another. All multicellular organisms produce hormones; plant hormones are also called phytohormones. Hormones in animals are often transported in the blood. Cells respond to a hormone when they express a specific receptor for that hormone. The hormone binds to the receptor protein, resulting in the activation of a signal transduction mechanism that ultimately leads to cell type-specific responses.

horticulture: It refers to gardening; growing fruits, vegetables, and ornamental plants.

host: In biology it refers to the organism that harbours another organism or parasite, or a mutual or commensal symbiont, typically providing nourishment and shelter. For example, a cell being host to a virus, a legume plant hosting helpful nitrogen-fixing bacteria, and animals as hosts to parasitic worms.

host-specific toxin: A metabolite produced by a pathogen which has a host specificity equivalent to that of the pathogen. Such toxins are utilized for in vitro selection experiments to screen for tolerance or resistance to the pathogen.

housekeeping genes: These are the genes that are always expressed (i.e., they are said to be constitutively expressed) due to their constant requirement by the cell.

hox genes: It refers to the genes that determine which proteins go where in a developing body.

Human Anti-Murine Antibody Response (HAMA): It is an immune response generated in humans to antibodies raised in murine (e.g., mouse or rat) cells.

humoral immune response: The immune response involving the transformation of B cells into plasma cells, which produce and secrete antibodies against a specific antigen.

humus: It refers to rich, black organic material; the living component of soils where plant and animal matter has been allowed to decompose.

hybrid arrested translation: A method for identifying the proteins coded by a cloned deoxyribonucleic acid (DNA) sequence by depending on the ability of the cloned DNA to form a base pair with its messenger ribonucleic acid and thereby inhibit its translation.

hybrid cell: Any cell type containing components from one or more genomes, other than zygotes and their derivatives. Hybrid cells may be formed by cell fusion or transfection. The mononucleate cell, which results from the fusion of two different cells, leads to the formation of a synkaryon.

hybrid dysgenesis: A syndrome of abnormal traits that appears in the hybrids between certain strains of the fruit fly *Drosophila melanogaster*, and includes such traits as partial sterility and greatly elevated genetic mutations and chromosome rearrangements. Its cause is mobilization of a transposable element.

hybrid released translation: A method used to detect the proteins encoded by cloned DNA. The cloned DNA is bound to a nitrocellulose filter and a crude preparation of mRNA is hybridized to the filter-bound DNA. Only mRNA sequences homologous to the cloned DNA will be retained on the filter. These mRNA molecules can then be removed by high temperature or by using formamide. The purified mRNA is then placed in an in vitro translation system and the proteins encoded by the message can be analysed by electrophoresis through a polyacrylamide gel.

hybrid: The offspring resulting from cross-breeding of different plants or animals. It is also in case of a heteroduplex DNA or DNA-RNA molecule.

hybridization: It is the interaction of complementary nucleic acid strands. This can occur between two DNA strands or DNA and RNA strands, and is the basis of many techniques such as Southern and Northern blots.

hybridization: It is the process of combining different varieties or species of organisms to create a hybrid. Even the pairing of two polynucleotide strands, often from different sources, by hydrogen bonding between complementary nucleotides.

hybridoma technology: It is a technology of forming hybrid cell lines by fusing a specific antibody-producing B cell with a myeloma (B cell cancer) cell that is selected for its ability to grow in tissue culture and an absence of antibody chain synthesis. The antibodies produced by the hybridoma are of a single specificity and are therefore monoclonal antibodies (in contrast to polyclonal antibodies). The production of monoclonal antibodies was invented by Cesar Milstein, Georges J.F. Köhler and Niels Kaj Jerne in 1975.

hydrocarbon: It refers to organic compounds composed primarily of hydrogen and carbon atoms. Petroleum, natural gas, coal, and methane are examples.

hydrologic cycle: It refers to the ongoing transformation of water in the biosphere from ocean water evaporation to clouds, rain, groundwater and run-off, storage in organisms, etc., until its return to the oceans. The earth holds roughly 326 cubic million miles of water, 97% of it is in the oceans.

hydrolysis: It is a chemical reaction during which the molecules of water (H_2O) are split into hydrogen cations (H^+) (conventionally referred to as protons) and hydroxide anions (OH^-).[1][2] It is a type of reaction that is used to break down certain polymers, especially those made by step-growth polymerization. Such polymer degradation is usually catalysed by either acid, e.g., concentrated sulphuric acid (H_2SO_4), or alkali, e.g., sodium hydroxide (NaOH) attack, often increasing with their strength or pH.

Hydrolysis is distinct from hydration. In hydration, the hydrated molecule does not "lyse" (break into two new compounds). It should not be confused with hydrogenolysis, a reaction of hydrogen.

hydrolysis: It refers to chemical weathering in which hydrogen ions from water alter a mineral's composition.

hydrophilicity: It refers to the degree to which a molecule is soluble in water. Hydrophilicity depends to a large degree on the charge and polarizability of the molecule and its ability to form transient hydrogen-bonds with (polar) water molecules.

hydrophobic interactions: Interaction of nonpolar (un-ionizable) hydrocarbon molecules forced together because of stronger water-water interaction.

hydrophobicity: It refers to the degree to which a molecule is insoluble in water, and hence is soluble in lipids. If a molecule lacking polar groups is placed in water, it will be entropically driven to finding a hydrophobic environment (such as the interior of a protein or a membrane).

hydroponics: It is a method of growing plants using mineral nutrient solutions, in water, without soil. Terrestrial plants may be grown with their roots in the mineral nutrient solution only or in an inert medium, such as perlite, gravel, or mineral wool.

hydroponics: It refers to growing plants without soil, usually by supplying nutrients to the roots by immersing them in a specially prepared solution or with special sprays (aeroponics).

hydrostatic pressure: It refers to the push of water under pressure, as in an aquifer.

hyperparasite: It refers to a parasite on another parasite. In crabs, pepper spot disease occurs when the flatworms that infect the crab are infected by a parasitic protozoan.

hyperploid: It is an individual, organism, strain, or cell that has one or more chromosomes in excess of the haploid number or an exact multiple of the haploid number characteristic of the species.

hypersensitive site: It is a short region of chromatin and is detected by its super sensitivity to cleavage by DNase I and other various nucleases (DNase II and micrococcal nucleases). In a hypersensitive site, the nucleosomal structure is not organized in the usual fashion, which results in a 100-fold increase in sensitivity to enzyme attack than in bulk chromatin.

hypersensitivity: It refers to undesirable (damaging, discomfort-producing and sometimes fatal) reactions produced by the normal immune system. Hypersensitivity reactions require a pre-sensitized (immune) state of the host.

hypertonic: It pertains to a solution (e.g., extracelllular fluid) with higher solute concentration compared with another. (see: hypotonic, isotonic). For example, if the extracellular fluid has greater amounts of solutes than the cytoplasm, the extracellular fluid is said to be hypertonic.

hypervariable region (HVR): It is a location within the nuclear DNA or the D-loop of mitochondrial DNA in which base pairs of nucleotides repeat (in the case of nuclear DNA) or have substitutions (in the case of mitochondrial DNA). Changes or repeats in the hypervariable region are highly polymorphic.

hypervariable segment: A region of a protein that varies considerably between strains or individuals.

hypochlorite: It is a chemical compound containing ClO^- group, with chlorine in oxidation state +1. Because of this oxidation state, the hypochlorite ion can be alternately named the chlorate(I) ion. Hypochlorites are salts of hypochlorous acid.

hypomorph: A mutation that causes a partial loss of gene function.

hypoplastic: Incomplete or underdevelopment of a tissue or organ. Hypoplastic left heart syndrome is the most serious type of congenital heart disease.

hypoploid: It is an individual, organism, strain, or cell that has fewer than the haploid number or an exact multiple of the haploid number of chromosomes characteristic of the species.

hypostasis: It is a change that occurs naturally after death in the human body; it is not an injury per se. In genetics, it refers to a condition in which the action of one gene conceals or suppresses the action of another gene that is not its allele, but affects the same part or biochemical process in an organism.

hypothesis: An assumption or speculation we make about a population parameter.

hypothesis: It is a proposed explanation for an observable phenomenon. Scientists generally base scientific hypotheses on previous observations that cannot be satisfactorily explained with the available scientific theories.

hypotonic: It refers to a solution with a comparatively lower concentration of solutes compared to another.

I

IAA: It is also known as Indole-3-acetic acid, and is a heterocyclic compound that is an phytohormone called auxins. This colourless solid is probably the most important plant auxin. The molecule is derived from indole, containing a carboxymethyl group (acetic acid).

ideogram: An ideogram or ideograph is a graphic symbol that represents an idea or concept. Some ideograms are comprehensible only by familiarity with prior convention; others convey their meaning through pictorial resemblance to a physical object, and thus may also be referred to as pictograms.

idiotype: It refers to the antibody variants localized to the variable portion of an immunoglobulin that are recognised by their antigenic determinants. The determinants are composed from the antigen-combining site or CDRs. Every unique antigenic determinant has a specific antibody with its own unique idiotype.

IGS: Internal guide sequence.

illumination: It is the deliberate application of light to achieve some aesthetic or practical effect. Lighting includes the use of both artificial light sources such as lamps and natural illumination of interiors from daylight.

illuviation: It refers to accumulation of soil materials leached from one soil layer or horizon into another.

imaginal disc: It is one of the parts of a holometabolous insect larva that will become a portion of the outside of the adult insect during the pupal transformation. Contained within the body of the larva, there are pairs of discs that will form, for instance, the wings or legs or antennae or other structures in the adult. During the pupal stage, many larval structures are broken down, and adult structures, including the discs, undergo rapid development. Each disc averts and elongates, with the central portion of the disc becoming the distal part of whichever appendage, wing, leg, antenna, etc., it is forming. During the larval stage, the cells in the growing disc appear undifferentiated, but their developmental fate in the adult is already determined.

imbibition: It is defined as the displacement of one fluid by another immiscible fluid. This process is controlled and affected by a variety of factors. The capillary number (*Ca*) and the mobility ratio (*M*) have the greatest importance. It is also defined as the phenomenon by which the living or dead plant cells absorb water by surface attraction.

immediate early genes (IEGs): These are the genes which are activated transiently and rapidly in response to a wide variety of cellular stimuli. They represent a standing response mechanism that is activated at the transcription level in the first round of response to stimuli, before any new proteins are synthesized. Thus IEGs are distinct from "late response" genes, which can only be activated later following the synthesis of early response gene products. Thus IEGs have been called the "gateway to the genomic response". The term can describe viral regulatory proteins that are synthesized following viral infection of a host cell, or cellular proteins that are made immediately following stimulation of a resting cell by extracellular signals.

immobilized cells: Cells entrapped in matrices such as alginate, polyacrylamide and agarose designed for use in membrane and filter bioreactors.

immortalization: To make cells, tissues or organisms immortal.

immortalizing oncogene: A gene that upon transfection enables a primary cell to grow indefinitely in culture.

immune response: The immune response is how your body recognizes and defends itself against bacteria, viruses, and substances that appear foreign and harmful to the body.

immune system: It is a system of biological structures and processes within an organism that protects against the diseases by identifying and killing pathogens and tumour cells. It detects a wide variety of agents, from viruses to parasitic worms, and distinguishes them from the organism's own healthy cells and tissues in order to function properly. Detection is complicated as pathogens can evolve rapidly, producing adaptations that avoid the immune system and allow the pathogens to infect their hosts successfully.

immunity: It is a biological term that describes a state of having sufficient biological defenses to avoid infection, disease, or other unwanted biological invasion. Immunity involves both specific and non-specific components. The non-specific components act either as barriers or eliminators of a wide range of pathogens irrespective of antigenic specificity. Other components of the immune system adapt themselves to the new disease encountered and are able to generate pathogen-specific immunity.

immunization: Immunization is the process whereby a person is made immune or resistant to a particular infectious disease, typically by the administration of a vaccine. Vaccines stimulate the body's own immune system to protect the person against subsequent infection or disease. It is the process by which an individual's immune system becomes fortified against an infectious agent.

immuno-affinity chromatography: A purification technique in which an antibody is bound to a matrix and is subsequently used to bind and separate a protein from a complex mixture. In this technique a specific antibody is covalently immobilised on a matrix, across which the sample to be purified is passed. The protein adsorbs to the ligand and can then be selectively desorbed in such a way as to retain the activity of the purified sample.

immunoassay: It is a biochemical test that measures the concentration of a substance in a biological liquid, typically serum or urine, using the reaction of an antibody with its antigen. The assay takes advantage of the specific binding of an antibody to its antigen. Monoclonal antibodies are often used as they usually bind to one site of a particular molecule, and therefore provide a more specific and accurate test, which is less easily confused by the presence of other molecules. The antibodies picked must have a high affinity for the antigen (if there is antigen available, a very high proportion of it must bind to the antibody).

immunodiagnostics: A diagnostic methodology that uses an antigen-antibody reaction as primary means of detection. The concept of using immunology as a diagnostic tool was introduced in 1960 as a test for serum insulin. A second test was developed in 1970 for thyroxine.

immunogenicity: It is the ability of a particular substance, such as an antigen or epitope, to provoke an immune response.

immunoglobulin: A protein produced by plasma cells and lymphocytes and characteristic of these types of cells. Immunoglobulins play an essential role in the body's immune system. They attach to foreign substances, such as bacteria, and assist in destroying them. Immunoglobulin is abbreviated as Ig. The classes of immunoglobulins are termed immunoglobulin A (IgA), immunoglobulin G (IgG), immunoglobulin M (IgM), immunoglobulin D (IgD) and immunoglobulin E (IgE).

immunosensor: Immunosensors transduce antigen-antibody interactions directly into physical signals. The design and preparation of an optimum interface between the biocomponents and the detector material is the key part of sensor development. An amperometric immunosensor-based on a new electrochemical detection scheme is under investigation.

immunosuppression: It involves an act that reduces the activation or efficacy of the immune system. Some portions of the immune system itself have immunosuppressive effects on other parts of the immune system, and immunosuppression may occur as an adverse reaction to the treatment of other conditions. Deliberately induced immunosuppression is generally done to prevent the body from rejecting an organ transplant, treating graft-versus-host disease after a bone marrow transplant, or for the treatment of autoimmune diseases such as rheumatoid arthritis or Crohn's disease. This is typically done using drugs, but may involve surgery (splenectomy), plasmapharesis, or radiation.

immunosuppressor: It is a drug that lowers the body's normal immune response.

immunotherapy: It is a medical term defined as "treatment of disease by inducing, enhancing, or suppressing an immune response."

immunotoxin: It is a human-made protein that consists of a targeting portion linked to a toxin. When the protein binds to that cell, it is taken in through endocytosis, and the toxin kills the cell. It is used for the treatment of some kinds of cancer and a few viral infections.

impeller: An impeller is a rotor inside a tube or conduit used to increase the pressure and flow of a fluid.

in silico (biology): It refers to the use of computers to simulate, process, or analyse a biological experiment.

in situ conservation: It means on-site conservation. It is the process of protecting an endangered plant or animal species in its natural habitat, either by protecting or cleaning up the habitat itself, or by defending the species from predators. This term also refers to the conservation of genetic resources in natural populations of plant or animal species, such as forest genetic resources in natural populations of tree species, and is increasingly being applied to conservation of agricultural biodiversity in agroecosystems by farmers, especially those using unconventional farming practices.

in situ hybridization (ISH): It is a type of hybridization that uses a labelled complementary DNA or RNA strand (i.e., probe) to localize a specific DNA or RNA sequence in a portion or section of tissue (in situ), or, if the tissue is small enough, in the entire tissue (whole mount ISH). This is distinct from immuno-histochemistry, which localizes proteins in tissue sections. DNA ISH can be used to determine the structure of chromosomes.

in situ hybridization: It refers to the variation of DNA/RNA hybridization procedure in which the denatured DNA is in place in the cell and is then challenged with RNA or DNA extracted from another source.

in situ: In Latin it means *in the natural place* or *in the original place.*

in vitro fertilization (IVF): It is a process by which egg cells are fertilised by sperm outside the womb, in vitro. IVF is a major treatment in infertility when other methods of assisted reproductive technology have failed. The process involves hormonally controlling the ovulatory process, removing ova (eggs) from the woman's ovaries and letting sperm fertilise them in a fluid medium. The fertilised egg (zygote) is then transferred to the patient's uterus with the intent to establish a successful pregnancy.

in vitro maturation (IVM): It is the technique of letting ovarian follicles mature in vitro.

in vitro transcription: It is a technique of cell-free transcription. The ability to synthesize RNA in the laboratory is critical to many techniques. Radiolabelled and non-isotopically labelled RNA probes, generated in small scale transcription reactions, can be used in blot hybridizations and nuclease protection assays. Such probes are much more sensitive than random-primed DNA probes. Small scale reactions may also be used to synthesize RNA transcripts containing modified nucleotides for various biochemical and molecular biology studies.

in vitro translation: It is a technique of cell-free translation. The in vitro synthesis of proteins in cell-free extracts is an important tool for molecular biologists and has a variety of applications, including the rapid identification of gene products, localization of mutations through synthesis of truncated gene products, protein folding studies, and incorporation of modified or unnatural amino acids for functional studies. The use of in vitro translation systems can have advantages over in vivo gene expression when the over-expressed product is toxic to the host cell, when the product is insoluble or forms inclusion bodies, or when the protein undergoes rapid proteolytic degradation by intracellular proteases.

in vivo gene therapy: One of the most amazing genetic applications in medicine is gene therapy. Also known as somatic gene therapy and therapeutic gene therapy, this procedure involves inserting (or sometimes deleting) portions of the genes in diseased patients so that they can be cured and live healthier lives.

in vivo: The natural conditions in which organisms reside.

inactivated agent: A virus, bacterium or other organism that has been treated to prevent it from causing a disease.

inbreeding depression: It is reduced fitness in a given population as a result of breeding of related individuals. Breeding between closely related individuals, called inbreeding, results in more recessive deleterious traits manifesting themselves. The more closely related the breeding pair is, the more homozygous deleterious genes the offspring may have, resulting in very unfit individuals. Another mechanism responsible is overdominance of heterozygous alleles leading to a reduction in the fitness of a population with many homozygous genotypes, even if they are not deleterious. Currently it is not known which of the two mechanisms is more important.

inbreeding: It is a genetic term that refers to reproduction as a result of mating of two animals which are genetically related to each other. If the relationship is a close one or practised repeatedly, inbreeding can increase the chances of offspring being affected by recessive or deleterious traits. This generally leads to a decreased fitness of a population, which is called inbreeding depression. Deleterious alleles causing inbreeding depression can subsequently be removed through culling, which is also known as genetic purging. The product of inbreeding is called inbred line.

incipient species: It refers to one which is about to become genetically isolated from its species of origin.

inclusion body: They are nuclear or cytoplasmic aggregates of stainable substances, usually proteins. They typically represent sites of viral multiplication in a bacterium or a eukaryotic cell and usually consist of viral capsid proteins. Inclusion bodies can also be hallmarks of genetic diseases, as in the case of neuronal inclusion bodies in disorders like frontotemporal dementia and Parkinson's disease.

incompatibility: In immunology, the degree to which the body's immune system will try to reject foreign material (as transfused blood or transplanted tissue).

incomplete dominance: Incomplete dominance is a form of intermediate inheritance in which one allele for a specific trait is not completely dominant over the other allele. This results in a combined phenotype.

incomplete penetrance: The frequency with which a heritable trait is manifested by individuals carrying the principal gene or genes conditioning it.

incubation: It is one of the 4 proposed stages of creativity: preparation, incubation, illumination, and verification. Incubation is defined as a process of unconscious recombination of thought elements that were stimulated through conscious work at one point in time, resulting in novel ideas at some later point in time. The experience of leaving a problem for a period of time, then finding the difficulty evaporates on returning to the problem, or even more striking, that the solution "comes out of the blue", when thinking about something else, is widespread. Many guides to effective thinking and problem solving advise the reader to set problems aside for a time.

incubator: An apparatus in which environmental conditions (light, photoperiod, temperature, humidity, etc.) are fully controlled, and used for hatching eggs and multiplying microorganisms.

independent assortment: Independent assortment is a basic principle of genetics developed by a monk named Gregor Mendel in the 1860's. Mendel formulated this principle after discovering another principle, now known as Mendel's law of segregation. This principle states that the alleles for a trait separate when gametes are formed. These allele pairs are then randomly united at fertilization. Mendel arrived at this conclusion by performing monohybrid crosses. These were cross-pollination experiments with pea plants that differed in one trait, for example, pod colour.

independent variable: The known variable or variables in regression analysis.

indicator plant: It refers to a plant whose presence and health give clues about the soil quality. Moss sometimes means a lot of acid in the soil, for example, although to a seasoned eye, most plants can serve as indicators.

indicator species: It refers to a species that works like an indicator plant: its population and health reveal much about its ecosystem.

indirect embryogenesis: Embryo formation on callus tissues derived from zygotic or somatic embryos, seedling plant or other tissues in culture.

indirect organogenesis: Organ formation on callus tissues derived from explants.

inducer: A low molecular weight compound or a physical agent that is bound by a repressor so as to produce a complex that can no longer bind to the operator. Thus, the presence of the inducer turns on the expression of the gene(s) controlled by the operator. For a gene to be expressed, its DNA sequence must be copied (in a process known as transcription) to make a smaller, mobile molecule called messenger RNA (mRNA), which carries the instructions for making a protein to the site where the protein is manufactured (in a process known as translation). Many different types of proteins can affect the level of gene expression by promoting or preventing transcription. In prokaryotes (such as bacteria), these proteins often act on a portion of DNA known as the operator at the beginning of the gene. The operator is where RNA polymerase, the enzyme which copies the genetic sequence and synthesizes the mRNA, attaches to the DNA strand.

inducible enzyme: An enzyme that is normally present in minute quantities within a cell, but whose concentration increases dramatically when a substrate compound is added.

inducible gene: A gene that is expressed only in the presence of a specific metabolite, the inducer.

induction media: Media used to induce the formation of organs or other structures.

induction: It refers to the initiation or cause of a change or process, such as the production of a specific morphogenetic effect in the developing embryo.

inembryonation: Non-surgical transfer of embryo(s) to a recipient female. As in vitro embryo technology develops, artificial inembryonation will gradually replace artificial insemination.

inert: The term inert is used to describe something that is not chemically active. The noble gases were described as being inert because they did not react with the other elements or themselves.

infauna: It refers to the animals that live in sedimentary deposits on the sea floor.

infection: It is the detrimental colonization of a host organism by a foreign species. In an infection, the infecting organism seeks to utilize the host's resources to multiply, usually at the expense of the host. The infecting organism, or pathogen, interferes with the normal functioning of the host and can lead to chronic wounds, gangrene, loss of an infected limb, and even death. The host's response to infection is inflammation. Colloquially, a pathogen is usually considered a microscopic organism, though the definition is broader, including parasites, fungi, viruses, prions, and viroids.

infectious agent: An agent capable of producing infection.

infiltration: It is the diffusion or accumulation (in a tissue or cells) of substances not normal to it or in amounts in excess of the normal. The material collected in those tissues or cells is also called infiltrate.

inheritance: It is the natural process in which the transfer of characters occurs from parents to offspring through genes during sexual reproduction. Genes are located on chromosomes. Chromosomes are in pairs and genes, or their alleles, are located on each of these pairs. The concept of inheritance was conclusively explained by G. J. Mendel in 1965 on the basis of simple experiments conducted on *Pisum sativum*.

inhibitor: Any substance or object that retards a chemical reaction; a major or modifier gene that interferes with a reaction or with the expression of another gene. Enzyme inhibitors are molecules that bind to enzymes and decrease their activity. Since blocking an enzyme's activity can kill a pathogen or correct a metabolic imbalance, many drugs are enzyme inhibitors. They are also used as herbicides and pesticides. Not all molecules that bind to enzymes are inhibitors; *enzyme activators* bind to enzymes and increase their enzymatic activity.

initiation codon (initiator): ATG and AUG denote sequences of DNA and RNA respectively that are the start codon or initiation codon encoding the amino acid methionine (Met) in eukaryotes and a modified Met (fMet) in prokaryotes.

initiation factors: They are the soluble proteins required for the initiation of translation. The start codon is typically AUG (or ATG in DNA; this also encodes methionine). Very rarely in higher organisms (eukaryotes) non AUG start codons used. In addition to AUG, alternative start codons, such as GUG and UUG are used in prokaryotes.

inoculation: It is the placement of something that will grow or reproduce, and is most commonly used in respect of introduction of a serum, vaccine, or antigenic substance into the body of a human or animal, especially to produce or boost immunity to a specific disease; but also can be used to refer to the communication of a disease to a living organism by transferring its causative agent into the organism, to implant microorganisms or infectious material into a culture medium such as a brewers vat or a Petri dish, to safeguard as if by inoculation, to introduce an idea or attitude into someone's mind, any placement of microorganisms or viruses at a site where infection is possible.

inoculum: It is the substance or small piece of tissue cut from callus, or an explant from a tissue or organ, or a small amount of cell material from a suspension culture, transferred into fresh medium for continued growth of the culture used in performing an inoculation.

inorganic compound: A chemical compound that is generally not derived from living processes; compounds that do not contain carbon.

inositol: It is also called cyclohexane-1,2,3,4,5,6-hexol is a chemical compound with formula $C_6H_{12}O_6$ or $(\text{-CHOH-})_6$, a sixfold alcohol (polyol) of cyclohexane. It exists in nine possible stereoisomers, of which the most prominent form, widely occurring in nature, is *cis*-1,2,3,5-*trans*-4,6-cyclohexanehexol, or *myo*-inositol. Inositol is a carbohydrate, though not a classical sugar. It is almost tasteless, with a small amount of sweetness.

insecticide: An insecticide is a pesticide used against insects. They include ovicides and larvicides used against the eggs and larvae of insects respectively. Insecticides are used in agriculture, medicine, industry and household.

insertion element: It is a section of DNA that is capable of becoming inserted into another chromosome.

insertion mutation: In genetics, it is the addition of one or more nucleotide base pairs into a DNA sequence. This can often happen in microsatellite regions due to the DNA polymerase slipping. Insertions can be anywhere in size from one base pair incorrectly inserted into a DNA sequence to a section of one chromosome inserted into another. On a chromosome level, an *insertion* refers to the insertion of a larger sequence into a chromosome. This can happen due to unequal crossover during meiosis.

insertion sequence (IS element): It is a short DNA sequence that acts as a simple transposable element. Insertion sequences have two major characteristics: they are small relative to other transposable elements (generally around 700 to 2500 bp in length) and only code for proteins implicated in the transposition activity (they are thus different from transposons, which also carry accessory genes such as antibiotic resistance genes). These proteins are usually the transposase, which catalyses the enzymatic reaction allowing the IS to move, and also one regulatory protein, which either stimulates or inhibits the transposition activity.

insertion site (cloning site): A unique restriction site in a vector DNA molecule into which foreign DNA can be inserted. The term is also used to describe the position of integration of a transposon or insertion site element.

insolation: It refers to the solar energy received by the earth.

instability (in systems): It is generally characterized by some of the outputs or internal states growing without bounds. Not all systems that are not stable are unstable; systems can also be marginally stable or exhibit limit cycle behaviour.

insulin: It is a hormone that has profound effects on metabolism. Insulin causes cells in the liver, muscle, and fat tissue to take up glucose from the blood, storing it as glycogen in the liver and muscle, and stopping use of fat as an energy source. When insulin is absent or low, glucose is not taken up by body cells, and the body begins to use fat as an energy source, for example, by transfer of lipids from adipose tissue to the liver for mobilization as an energy source.

As its level is a central metabolic control mechanism, its status is also used as a control signal to other body systems.

integrating vector: A vector that is designed to integrate cloned DNA into the host cell chromosomal DNA.

integration: It refers to the physical insertion of DNA into the host cell genome. The process is used by retroviruses where a specific enzyme catalyses the process or can occur at random sites with other DNA.

integration: The recombination process which inserts a small DNA molecule (usually by homologous recombination) into a larger one. If the molecules are circular, integration involves only a single crossing-over; if linear, then two crossings-over are required.

integration-excision (I/E) region: The portion of bacteriophage lambda DNA that enables bacteriophage lambda DNA to be inserted into a specific site in the *E. coli* chromosome and excised from this site.

intensifying screen: It is a sheet of crystals of inorganic salts (called phosphors) which emit fluorescent light when excited by X-ray radiation. The sheets are used to intensify the effect of X-rays during exposure of X-ray film.

interaction: It is a kind of action that occurs as two or more objects have an effect upon one another. The idea of a two-way effect is essential in the concept of interaction, as opposed to a one-way causal effect. A closely related term is interconnectivity, which deals with the interactions of interactions within systems: combinations of many simple interactions can lead to surprising emergent phenomena. *Interaction* has different tailored meanings in various sciences. All systems are related and interdependent. Every action has a consequence.

intercalating agent: A chemical substance that can insert itself between the base pairs in a deoxyribonucleic acid molecule.

interference (statistical): When two probability distributions overlap, statistical interference exists. Knowledge of the distributions can be used to determine the likelihood that one parameter exceeds another, and by how much.

interference: Crossing over at one point that alters the chance of another crossing-over nearby; detected by studying the pattern of crossings-over with three or more linked genes. Interference is positive or negative depending on whether the chance of another crossing-over nearby is reduced or increased, respectively.

interferons (IFNs): They are proteins made and released by the cells of most vertebrates in response to the presence of pathogens, such as viruses, bacteria, or parasites, or tumour cells. They allow communication between the cells to trigger the protective defences of the immune system that eradicate pathogens or tumours.

intergenic region (IGR): It is a stretch of DNA sequences located between the clusters of genes that contain few or no genes. Occasionally, some intergenic DNA acts to control genes close by, but most of it has no currently known function. It is one of the DNA sequences collectively referred to as junk DNA, though it is only one phenomenon labelled such and in scientific studies today, the term is less used. In humans, intergenic regions comprise a large percentage of the genome.

interleukins: They are a group of cytokines (secreted proteins/signalling molecules) that were first seen to be expressed by white blood cells (leukocytes). The function of the immune system depends in a large part on *interleukins*, and rare deficiencies of a number of them have been described, all featuring autoimmune diseases or immune deficiency.

interphase: It is the phase of the cell cycle in which the cell spends the majority of its time and performs the majority of its purposes including preparation for cell division. In preparation for cell division it increases its size and number of organelles, and makes a copy of its DNA. Interphase is also considered to be the 'living' phase of the cell, in which the cell obtains nutrients, grows, reads its DNA, and conducts other "normal" cell functions. The majority of eukaryotic cells spend most of their time in interphase. Interphase does not describe a cell that is merely resting but is rather an active preparation for cell division. A common misconception is that interphase is the first stage of mitosis, however, since mitosis is the division of the nucleus, prophase is actually the first stage.

interspecific: Between two different species.

intracellular signalling: It refers to the communication of a molecular message from the surface of the cell to the nucleus via the participation of a series of molecules, including receptors, enzymes, proteins, and small molecules. The end result of the signalling process is the up- or down-regulation of a particular series of genes that may be involved in cell growth, division or differentiation.

intracellular: Occurring within a cell.

intracytoplasmic sperm injection (ICSI): It is an in vitro fertilization procedure in which a single sperm is injected directly into an egg. This procedure is most commonly used to overcome male infertility problems, although it may also be used where eggs cannot be easily penetrated by sperm, and occasionally as a method of in vitro fertilization, especially that associated with sperm donation.

intragenic complementation: Complementation that occurs between two mutant alleles of a gene; common only when the product of the gene functions as a homomultimer.

intraspecific: Within a species or its populations.

introgression (introgressive hybridization): It is the movement of a gene from one species into the gene pool of another by repeated backcrossing an interspecific hybrid with one of its parent species. Introgression is a long-term process; it may take many hybrid generations before the backcrossing occurs.

intron (intervening sequence): It is a DNA region within a gene that is not translated into protein. These non-coding sections are transcribed to precursor mRNA (pre-mRNA) and some other RNAs (such as long noncoding RNAs), and subsequently removed by a process called splicing during the processing to mature RNA. After intron splicing (i.e., removal), the mRNA consists only of exon derived sequences, which are translated into a protein. The word *intron* is derived from the term *intragenic region* and is also called *intervening sequence* (IVS).

invariant: Anything unchanging, usually referring to the portion of a molecule that is the same across species.

invasiveness: Ability of a plant to spread beyond its introduction site and become established in new locations where it may have a deleterious effect on organisms already existing there.

inverse relationship: A relationship between two variables in which as the independent variable increases, the dependable variable decreases.

inversion: A type of chromosomal aberration, wherein re-arrangement chromosome reverses the order of a linear array of genes in it.

inverted repeat (IR): It is a sequence of nucleotides that is the reversed complement of another sequence further downstream. For example, 5'—GACTGC.... GCAGTC—3'. When no nucleotides intervene between the sequence and its downstream complement, it is called a palindrome. Inverted repeats define the boundaries in transposons. Inverted repeats also indicate the regions capable of self-complementary base pairing.

ion channels: They are pore-forming proteins that help establish and control the small voltage gradient across the plasma membrane of all living cells (see cell potential) by allowing the flow of ions down their electrochemical gradient. They are present in the membranes that surround all biological cells. The study of ion channels is known as channelomics and involves many scientific techniques such as voltage clamp electrophysiology (in particular patch clamp), immunohistochemistry, and RT-PCR.

ion: A charged particle.

ionic bonds: Attractions between oppositely charged chemical groups.

ionizing radiation: It consists of subatomic particles or electromagnetic waves that are energetic enough to detach electrons from atoms or molecules, ionizing them. The occurrence of ionization depends on the energy of the impinging individual particles or waves, and not on their number.

IPTG (Isopropyl-D-thiogalactopyranoside): It is an inducer of the *lac* (lactose) operon. In recombinant DNA technology, IPTG is often used to induce cloned genes that are under the control of *lac* repressor-*lac* promoter system.

irradiation: It is the process by which an item is exposed to radiation. The exposure can be intentional, sometimes to serve a specific purpose, or it can be accidental. In common usage the term refers specifically to ionizing radiation, and to a level of radiation that will serve that specific purpose, rather than radiation exposure to normal levels of background radiation or abnormal levels of radiation due to accidental exposure. This term also applies to 'non-ionizing radiation (as microwaves) or low frequency (50/60 Hz power supply), high frequency (as cellular phones, radio and TV transmissions).

IS element (insertion sequence): It is a short DNA sequence that acts as a simple transposable element. Insertion sequences have two major characteristics: they are small relative to other transposable elements (generally around 700 to 2500 bp in length) and only code for proteins implicated in the transposition activity (they are thus different from other transposons, which also carry accessory genes such as antibiotic resistance genes).

isoalleles: Different forms of a gene that produce the same phenotype or very similar phenotypes.

isochromosome: It is a chromosome that has lost one of its arms and replaced it with an exact copy of the other arm. This is sometimes seen in some females with Turner syndrome or in tumour cells.

isodiametric: Term commonly used to describe cells with equal diameters.

isoelectric focusing (IEF): It is a technique for separating different molecules by their electric charge differences. It is a type of zone electrophoresis, usually performed in a gel, that takes advantage of the fact that a molecule's charge changes with the pH of its surroundings. A protein that is in a pH region below its isoelectric point (pI) will be positively charged and so will migrate towards the cathode. As it migrates through a gradient of increasing pH, however, the protein's overall charge will decrease until the protein reaches the pH region that corresponds to its pI. At this point it has no net charge and so migration ceases. As a result, the proteins become focused into sharp stationary bands with each protein positioned at a point in the pH gradient corresponding to its pI. The technique is capable of extremely high resolution with proteins differing by a single charge being fractionated into separate bands.

isoenzyme: They are proteins that catalyze chemical reactions. In enzymatic reactions, the molecules at the beginning of the process are called substrates, and the enzyme converts them into different molecules, called the products. Almost all processes in a biological cell need enzymes to occur at significant rates. Since enzymes are selective for their substrates and speed up only a few reactions from among many possibilities, the set of enzymes made in a cell determines which metabolic pathways occur in that cell.

isoform: Any of the several different forms of the same protein, arising from either single nucleotide.

isogamy: Sexual reproduction involving the fusion of gametes that are similar in size and structure.

isogenic stocks: Strains of organisms that are genetically identical and completely homozygous.

isolation medium: An optimum nutrient medium suitable for the survival, growth and development of explants.

isoleucine (Ile or I): It is an α-amino acid with the chemical formula $HO_2CCH(NH_2)CH(CH_3)CH_2CH_3$. It is an essential amino acid, which means that humans cannot synthesize it, so it must be ingested. Its codons are AUU, AUC and AUA.

isomerise: It is an enzyme that catalyzes the structural rearrangement within a single olecule. It is also said to catalyze the conversion of one isomer into another.

Isomerases have their own EC classification of enzymes, **EC 5**. Isomerases can be further classified into six subclasses.

iso-osmotic (isotonic): Solutions having the same total osmotic pressure or osmolality as another fluid (ordinarily intracellular fluid); such a fluid is not iso-osmotic if it includes solutes that freely permeate cell membranes.

isoschizomers: It refers to the two different restriction enzymes which recognize and cut DNA at the same recognition site.

isotope: They are different types of atoms (nuclides) of the same chemical element, each having a different number of neutrons. Correspondingly, isotopes differ in mass number (or number of nucleons), but not in atomic number. The number of protons (the atomic number) is the same because that is what characterizes a chemical element.

isozyme (isoenzymes): These are enzymes that differ in amino acid sequence but catalyze the same chemical reaction.

isozymes: It refers to the two or more enzymes capable of catalyzing the same reaction but varying in their specificity due to differences in their structures and hence their efficiencies under different environmental conditions.

iteration: It refers to a series of steps in an algorithm whereby the processing of data is performed repetitively until the result exceeds a particular threshold. Iteration is often used in multiple sequence alignments whereby each set of pairwise alignments are compared with every other, starting with the most similar pairs and progressing to the least similar, until there are no longer any sequence-pairs remaining to be aligned.

J

joint probability: The probability of two events occurring together.

judgement sampling: In judgement sampling, the choice of sample items depends exclusively on the judgement of the investigator.

jumping genes (transposable elements): They are sequences of DNA that can move around to different positions within the genome of a single cell, a process called **transposition**. In the process, they can cause mutations and change the amount of DNA in the genome.

junk DNA: It refers to the term used to describe the excess DNA that is present in the genome beyond that required to encode proteins. A misleading term, since these regions are likely to be involved in gene regulation, and other as yet unidentified functions.

juvenile hormone: They are a group of acyclic sesquiterpenoids that regulate many aspects of insect physiology.

K

kanamycin (kanamycin sulfate): It is an aminoglycoside antibiotic, available in both oral and intravenous forms, and used to treat a wide variety of infections. Kanamycin is isolated from *Streptomyces kanamyceticus*.

kanr: Kanamycin-resistance gene.

kappa chain: Either of two types of light chains (small polypeptide chains) that are found in immunoglobulins.

karyogamy: It is the fusion of pronuclei of two cells, as part of syngamy, fertilization, or true bacterial conjugation. It is one of the two major modes of reproduction in fungi. In fungi that lack sexual cycles, it is an important source of genetic variation through the formation of somatic diploids.

karyokinesis: During cell division, the process of partition of a cell's nucleus into the daughter cells.

karyotype: A karyotype is the characteristic chromosome complement of a eukaryote species. The term may also be used for a cell, or an individual.

Kb: It refers to kilobase and is a unit of measurement in molecular biology equal to 1000 base pairs of DNA or RNA. **Kilobase** (kb) a length unit equal to 1000 base pairs of a double-stranded nucleic acid molecule. One kilobase of double-stranded DNA has a mass of about 660 kilodalton.

Kcat: It refers to the overall catalytic rate of an enzyme, or the maximum number of enzymatic reactions catalyzed per second.

keystone species: It refers to a species that increases or decreases the diversity of a system. For example, otters, which when hunted to extinction, remove a check on the sea urchins that will eat local kelp forests on which many other species depend.

killer T cells (Natural killer T cells): They are a heterogeneous group of T cells that share properties of both T cells and natural killer (NK) cells. Many of these cells recognize the non-polymorphic CD1d molecule, an antigen-presenting molecule that binds self- and foreign lipids and glycolipids. They constitute only 0.2% of all peripheral blood T cells.

kilocalorie (kcal): It is equal to 1000 cal.

kilodalton (kDa): A unit of atomic mass equal to 1000 daltons.

kinase: It is also known as a phosphotransferase. It is a type of enzyme that transfers phosphate groups from high-energy donor molecules, such as ATP, to specific substrates. The process is referred to as phosphorylation. An enzyme that removes phosphate groups is known as a phosphatase.

kinetics: It is a term for the branch of classical mechanics that is concerned with the relationship between the motion of bodies and its causes, namely, forces and torques.

kinetin: It is a kind of cytokinin, a class of plant hormone that promotes cell division. It was given the name kinetin because of its ability to induce cell division, provided that auxin was present in the medium.

kinetochore: It is the protein structure on chromosomes where the spindle fibres attach during division to pull the chromosomes apart. The kinetochore forms in eukaryotes and assembles on the centromere and links the chromosome to microtubule polymers from the mitotic spindle during mitosis and meiosis.

kinetosome (basal body, blepharoplast): It is an organelle formed from a centriole. It is a cylindrical structure in the cytoplasm from which a flagellum or cilium arises in motile eukaryotic cells. It is composed of a cylinder of nine microtubules linked by fine crossbridges. It transfers ATP and possibly other materials to the fibres in the shaft of the flagellum or cilium. Striated fibrous extensions, termed rootlet fibres, may arise from the basal body and project backwards into the cytoplasm. These probably serve to anchor the flagellum.

kinin: It is any of various structurally related polypeptides, such as bradykinin and kallikrein. They are members of the autacoid family.

kinship gardening: It refers to a form of companion planting in which botanical kin are placed next to each other to foster mutual growth and health.

Klenow fragment: It is a large protein fragment produced when DNA polymerase I from *E. coli* is enzymatically cleaved by the protease subtilisin.

Km: The catalytic efficiency of an enzyme-catalysed reaction. The greater the value of **kcat/Km,** the more rapidly and efficiently the substrate is converted into product.

knockout mice (gene targeting): It refers to the mice which have been engineered to lack a chosen gene. The gene is inactivated in so-called embryonic stem cells using the technique of homologous recombination. These cells are then introduced into a early stage embryo (blastocyst) and this is then transplanted into a recipient mouse. The subsequent progeny lack the targeted gene in some cells. This technique is used to determine the function of the chosen gene.

Krebs cycle: It refers to the second stage in the breakdown of glucose (unprocessed sugar) into energy (ATP), carbon dioxide (as waste), and water. Citric acid entering the Krebs cycle as a product of glycolysis (first stage) exits as NADH (reduced nicotinamide adenine dinucleotide), one molecule of GTP (guanosine triphosphate), and one molecule of $FADH_2$ (reduced lavin adenine dinucleotide). NADH and $FADH_2$ give their high-energy electrons to an electron transport system (third stage) that makes many ATP molecules.

kurtosis: The degree of peakedness of a distribution.

L

lab on a chip: It refers to the term describing microdevices that allow rapid, microanalytical analysis of DNA or protein in a single, fully integrated system. Typically, these devices are miniature surfaces, made of silicon, glass or plastic, which carry the necessary microdevices (pumps, valves, microfluidic controllers, and detectors) that allow sample separation and analysis. These devices are used in drug discovery, genetic testing and separation science.

label: A compound or atom that is either attached to or incorporated into a macromolecule and is used to detect the presence of a compound, substance, or macromolecule in a sample. The process of replacing a stable atom in a compound with a radioactive isotope of the same element to enable it to be detected by autoradiography or other techniques is called labelling. Increasingly, radioactive labelling is being replaced by fluorescent labelling.

***lac* repressor:** It is a DNA-binding protein which inhibits the expression of genes coding for proteins involved in the metabolism of lactose in bacteria. It is active in the absence of lactose, ensuring that the bacterium only invests energy in the production of machinery necessary for the uptake and metabolism of lactose when lactose is present. When lactose becomes available, it is converted into allolactose, which inhibits the *lac* repressor's DNA binding ability.

lactose: It is a sugar that is found most notably in milk. Lactose makes up around 2–8% of milk (by weight), although the amount varies among species and individuals. It is extracted from sweet or sour whey. The name comes from *lacte*, the Latin word for milk, plus the *-ose* ending used to name sugars.

lag phase: The initial growth phase, during which cell number remains relatively constant prior to rapid growth.

lagging strand: It is that template strand of DNA double helix that is oriented in a 5' to 3' manner. The newly made strand will be antiparallel, oriented in the 3' to 5' direction. Because of its orientation, the newly made strand is extending opposite to the working orientation of DNA polymerase III which is in a 5' to 3' manner. As a result, replication of the lagging strand is more complicated than that of the leading strand.

lambda chain: Although light chains are found in many multimeric proteins, λ chain usually refers to the light chains of immunoglobulins. These are of 22 kd and one of the two types, kappa or lambda. A single immunoglobulin has identical light chains (2 kappa or 2 lambda). Light chains have one variable and one constant region. There are isotype variants of both kappa and lambda.

lamella: It is a term for a plate-like structure, appearing in multiples, that occurs in various situations, such as biology or materials sciences. It implies a thin layer (Latin), the same derivation as for 'laminate'.

lamina: The flat part of a leaf, an organ of a plant.

laminar air-flow cabinet (tissue culture hood): It is a carefully enclosed bench designed to prevent contamination of semiconductor wafers, biological samples, or any particle sensitive device. Air is drawn through a HEPA filter and blown in a very smooth, laminar flow towards the user. The cabinet is usually made of stainless steel with no gaps or joints where spores might collect.

laminarin: It is a storage polysaccharide of glucose found in brown algae. It is used as a carbohydrate food reserve in the same way that chrysolaminarin is used by phytoplankton.

lampbrush chromosome: It is the largest chromosome known and is found in the amphibian oocytes (immature eggs). Lampbrush chromosomes occur during the diplotene stage of meiosis I. They are meiotic bivalents, each consisting of 2 sister chromatids. Lampbrush chromosomes are clearly visible even in the light microscope, where they are seen to be organized into a series of large chromatin loops emanating from a linear chromosomal axis.

landrace: It refers to domesticated animals or plants adapted to the natural and cultural environment in which they live. It differs from a formal breed which has been selectively bred deliberately to conform to a particular formal, purebreed standard of traits. Landraces are usually more genetically and physically diverse than formal breeds.

larva: It refers to a recently born or hatched animal before it undergoes metamorphosis into its mature form.

latent agent: It is usually a virus that is present in a host organism without producing any symptoms.

latent bud: An axillary bud whose development is inhibited, sometimes for many years, due to the influence of apical and other buds.

LCR (ligase chain reaction): It is a method of DNA amplification. While the better-known PCR carries out the amplification by polymerizing nucleotides, LCR instead amplifies the nucleic acid used as the probe. For each of the two DNA strands, two partial probes are ligated to form the actual one; thus, LCR uses two enzymes: a DNA polymerase and a DNA ligase. Each cycle results in a doubling of the target nucleic acid molecule. A key advantage of LCR is greater specificity as compared to PCR.

LD50: It is an abbreviation for "Lethal Dose, 50%" or median lethal dose.

leaching: It refers to the flushing or percolation of chemicals, minerals, or other substances through soil. Pesticides, fertilizers, poisons from mines or feedlots, and wastes from industrial plants sometimes leach into groundwater. Leaching also refers to washing the salt from soil to increase its fertility.

lead compound: It refers to the candidate compound identified as the best "hit" (tight binder) after screening of a combinatorial (or other) compound library, which is then taken into further rounds of screening to determine its suitability as a drug.

lead optimization: It refers to the process of converting a putative lead compound ("hit") into a therapeutic drug with maximal activity and minimal side affects, typically using a combination of computer-based drug design, medicinal chemistry and pharmacology.

leader peptide: Short peptide sequences translated from bacterial leader RNA sequences which are involved in translation attenuation, a mechanism that modulates mRNA translation.

leader sequence: It is also called five prime untranslated region (5' UTR). It is a particular section of messenger RNA (mRNA) and the DNA that codes for it. It starts at the +1 position (where transcription begins) and ends just before the start codon (usually AUG) of the coding region.

leading strand: It is the strand of DNA being replicated continuously. In DNA replication, the strand that is made in the 5' to 3' direction by continuous polymerization at the 3' growing tip.

lectin: Lectins are sugar-binding proteins which are highly specific for their sugar moieties. They typically play a role in biological recognition phenomena involving cells and proteins. For example, some viruses use lectins to attach themselves to the cells of the host organism during infection.

legume: It refers to a flowering bean or pea plant. All legumes contribute to the nitrogen, phosphorus (key to flowering and seed formation), and potassium nutrition of the crops that follow. Alfalfa is the heaviest nitrogen producer.

leptokurtic: A strongly peaked distribution.

leptonema: It refers to the chromosomes at the first stage of prophase-1 of meiosis.

lethal gene: A gene which brings about the death of the organism carrying it. Lethal genes constitute the most common class of mutations and are reflections of the fact that the fundamental function of genes is the control of processes essential to the growth and development of organisms.

lethal mutation: A type of mutation in which the effect(s) can result in the death or reduce significantly the expected longevity of an organism carrying the mutation.

leucine zipper: It refers to the protein motif which binds DNA in which 4-5 Leucines are found at 7 amino acid intervals. This motif is present typically in transcription factors and other proteins that bind DNA.

lexicon: In bioinformatics, a lexicon refers to a predefined list of terms that together completely define the contents of a particular database (strict). The component in the grammar which is in bare form a list of words or lexical entries.

lichen: It refers to the symbiotic association of a true fungi and a green alga. The alga makes sugar through photosynthesis, and the fungi fills out the organism while holding it to a surface.

life cycle: It is a period involving all different generations of a species succeeding each other by means of asexual or sexual reproduction. It also refers to the course of developmental changes through which an organism passes from its inception as a fertilized zygote to the mature state in which another zygote may be produced.

ligand: A ligand is a substance that is able to bind to and form a complex with a biomolecule to serve a biological purpose. In a narrower sense, it is a signal triggering molecule, binding to a site on a target protein.

ligase: It is an enzyme that can catalyse the joining of two large molecules by forming a new chemical bond, usually with accompanying hydrolysis of a small chemical group pendant to one of the larger molecules.

ligation: The process of joining two pieces of DNA to a single piece through the use of ligase.

lignification: A change in the character of a cell wall, by which it becomes harder. It is supposed to be due to an incrustation of lignin.

lignin: It is a complex chemical compound most commonly derived from wood, and an integral part of the secondary cell walls of plants and some algae.

lignocellulosic biomass: It refers to plant biomass that is composed of cellulose, hemicellulose, and lignin.

line diagram: Straight lines whose length is proportional to the frequencies.

lineage (evolutionary): It is a sequence of species that forms a line of descent; each new species is the direct result of speciation from an immediate ancestral species.

linear phase: It refers to a filter or other network whose image phase constant is a linear function of frequency. In a causal system, perfect linear phase can be achieved with a discrete-time FIR filter. Linear phase system has the property of the true time delay. Systems with generalized linear phase have an additional frequency-independent constant added to the phase. Because of this constant, the phase of the system is not a strictly linear function of frequency, but it retains many of the useful properties of linear phase systems.

linear regression analysis: The statistical analysis employed to find out the exact position of the straight line or lines is known as the linear regression analysis.

linear relationship: A particular type of association between two variables that can be described statistically by a straight line.

LINEs (long interspersed nuclear elements): They are a group of genetic elements that are found in large numbers in eukaryotic genomes.

linkage (genetic): It occurs when particular genetic loci or alleles for genes are inherited jointly. Genetic loci on the same chromosome are physically close to one another and tend to stay together during meiosis, and are thus genetically linked. This is called autosomal linkage. Alleles for genes on different chromosomes are usually not linked, due to independent assortment of chromosomes during meiosis.

linkage disequilibrium: It is the non-random association of alleles at two or more loci, not necessarily on the same chromosome. It is not the same as linkage, which describes the association of two or more loci on a chromosome with limited recombination between them. Linkage disequilibrium describes a situation in which some combinations of alleles or genetic markers occur more or less frequently in a population than would be expected from a random formation of haplotypes from alleles based on their frequencies.

linkage equilibrium: When the observed frequencies of haplotypes in a population agrees with haplotype frequencies predicted by multiplying together the frequency of individual genetic markers in each haplotype.

linkage map: A linkage map is a genetic map of a species or experimental population that shows the position of its known genes or genetic markers relative to each other in terms of recombination frequency, rather than as specific physical distance along each chromosome. Genetic maps help researchers to locate other markers, such as other genes by testing for genetic linkage of the already known markers.

linked genes: Genes that are inherited together with the other gene(s) as they are located on the same chromosome.

linker (DNA): It is a short segment of DNA with many restriction sites; commonly used in biotechnology, bioengineering, and molecular genetics.

lipases: It is a water-soluble enzyme that catalyzes the hydrolysis of ester bonds in water-insoluble, lipid substrates. Lipases thus comprise a subclass of the esterases.

lipid: Lipids are a broad group of naturally-occurring molecules which include fats, waxes, sterols, fat-soluble vitamins (such as vitamins A, D, E and K), monoglycerides, diglycerides, phospholipids, and others. The main biological functions of lipids include energy storage, as structural components of cell membranes, and as important signalling molecules.

lipofection (liposome transfection): It is a technique used to inject genetic material into a cell by means of liposomes, which are vesicles that can easily merge with the cell membrane since they both are made of a phospholipid bilayer. Lipofection is a lipid-based transfection technology which belongs to biochemical methods including polymers, DEAE dextran and calcium phosphate.

lipopolysaccharide (LPS): These are also known as lipoglycans, and are large molecules consisting of a lipid and a polysaccharide joined by a covalent bond. They are found in the outer membrane of Gram-negative bacteria. They act as endotoxins and elicit strong immune responses in animals.

liposome: It is a tiny bubble (vesicle) made of the same material as a cell membrane. Liposomes can be filled with drugs, and are used to deliver drugs for cancer and other diseases. Membranes are usually made of phospholipids, which are molecules that have a head group and a tail group. The head is attracted to water, and the tail, which is made of a long hydrocarbon chain, is repelled by water.

liquefaction: It is a change from gas to liquid through condensation, usually by cooling, or a change from solid to liquid through melting, usually by heating, or by grinding and blending with another liquid to induce dissolution. Earthquakes can also cause liquefaction.

liquid media: It refers to the nutrient media without any solidifying agent (agar). It is used in the culturing of cells and plant protoplasts.

liquid membranes: It refers to a membrane made of liquid.

liquid nitrogen: It is nitrogen in a liquid state at a very low temperature.

liquidators: It is the name given in the former USSR to approximately 8,00,000 people who were in charge of the removal of the consequences of the April 26, 1986 Chernobyl disaster on the site of the event.

litmus: It is a water-soluble mixture of different dyes extracted from lichens, especially *Roccella tinctoria*. It is often absorbed onto filter paper. The resulting piece of paper or solution with water becomes a pH indicator (one of the oldest), which is used to test materials for acidity.

live vaccine: It is a vaccine prepared from live microorganisms that have been attenuated but retain their immunogenic properties.

locus of control: It refers to a person's belief about what causes the good or bad results in his life, either in general or in a specific area such as health or academics.

locus: It is a specific location of a gene or DNA sequence on a chromosome. A variant of the DNA sequence at a given locus is called an allele.

locust bean gum (carob gum, carob bean gum, carobin or E410): It is a galactomannan vegetable gum extracted from the seeds of Carob tree. It is used as a thickening and gelling agent in food technology. It is soluble in hot water.

lod score: It is a statistical estimate of whether two loci are likely to lie near each other on a chromosome and are therefore likely to be inherited together as a package. LOD stands for logarithm of the odds (to the base 10).

log phase (logarithmic phase or exponential growth phase): It is a period characterized by cell doubling. The number of new bacteria appearing per unit time is proportional to the present population. If the growth is not limited, doubling will continue at a constant rate, resulting both the number of cells and the rate of population increasing doubly with each consecutive time period.

long template: A DNA strand that is synthesized during the polymerase chain reaction and has a primer sequence at one end but is extended beyond the site that is complementary to the second primer at the other end.

long terminal repeat (LTR): They are sequences of DNA that repeat hundreds or thousands of times. They are found in retroviral DNA and retrotransposons, flanking functional genes. They are used by viruses to insert their genetic sequences into the host genomes.

long-day plant: It refers to the plants classified based on photoperiodism**.** Long-day plant requires fewer than a certain number of hours of darkness in each 24-hour period to induce flowering. These plants typically flower in the northern hemisphere during late spring or early summer as days get longer.

loop bioreactors: They are the bioreactors in which the fermenting material is cycled between a bulk tank and a smaller tank or loop of pipes. The circulation helps mix the materials and ensures that gas injected into the fermenter is well distributed.

lower-tailed test: A one tailed hypothesis test in which a sample value significantly below the hypothesized population value will lead us to reject the null hypothesis.

LTR retrotransposons: They are the transposons ranging from ~100 bp to over 5 kb in size. Genome, and approximately 10% of the mouse genome are composed of the LTR transposons.

luteinizing hormone (LH) or lutropin: It is a hormone produced by the anterior pituitary gland.

lux (lx): It is the SI unit of illuminance and luminous emittance. It is used in photometry as a measure of intensity, as perceived by the human eye, of light that hits or passes through a surface.

luxury consumption: Nutrient absorption by an organism in excess of that required for optimum growth and productivity.

lyase: It is an enzyme that catalyzes the breaking of various chemical bonds by means other than hydrolysis and oxidation, often forming a new double bond or a new ring structure.

lymphocyte: It is a type of white blood cell in the vertebrate immune system.

lymphocytosis: It is an increase in the number or proportion of lymphocytes in the blood, usually detected when a complete blood count is routinely obtained. Lymphocytes normally represent 20 to 40% of circulating white blood cells.

lymphokine: It is a subset of cytokines that are produced by a type of immune cell known as a lymphocyte. They are typically produced by T cells to direct the immune system response by signalling between its cells. Lymphokines have many roles, including the attraction of other immune cells, like macrophages and other lymphocytes, to an infected site, and their subsequent activation prepares them to attack the invaders. Circulating lymphocytes can detect a very small concentration of lymphokine and then move up the concentration gradient towards where the immune response is required.

lymphokine-activated killer cell (LAK cell): It is a white blood cell that has been stimulated to kill tumour cells.

lymphoma: It is a cancer that begins in the lymphocytes of immune system and presents as a solid tumour of lymphoid cells. It is treatable with chemotherapy, and in some cases radiotherapy and/or bone marrow transplantation, and is curable, depending on the histology, type, and stage of the disease.

lyophilize: It refers to a process of freeze-drying. The material during lyophilization is rapidly frozen and dehydrated under high vacuum.

lysate: A fluid containing the contents of lysed cells is called a lysate.

lysis: It refers to the death of a cell by breaking of the cellular membrane, causing the contents to spill out, often by viral, enzymic or osmotic mechanisms that compromise its integrity.

lysogen: It is a bacteria that has been infected by any number of phages.

lysogenic cycle: It is one of two phases of viral reproduction. It is characterized by the integration of bacteriophage nucleic acid into the host bacterium's genome.

lysogeny: The fusion of nucleic acid of a bacteriophage with that of a host bacterium so that the potential exists for the newly integrated genetic material to be transmitted to daughter cells during each subsequent cell division.

lysosome: It is spherical organelle that contains enzymes (acid hydrolases). They break up food so that it is easier to digest. They are found in animal cells, while in yeast and plants the same roles are performed by lytic vacuoles.

lytic cycle: It is one of the two cycles of viral reproduction, the other being the lysogenic cycle. The lytic cycle is typically considered the main method of viral replication, since it results in the destruction of infected cell.

M

M 13: It is a filamentous bacteriophage composed of circular single-stranded DNA (ssDNA), which is 6407 nucleotides long, encapsulated in approximately 2700 copies of the major coat protein P8, and capped with 5 copies of two different minor coat proteins (P9, P6, P3) on the ends.

maceration: It is a technique to extract organic microfossils from a surrounding rock matrix using acid. Hydrochloric acid or acetic acid may be used to extract phosphatic fossils, such as the small shelly fossils, from a carbonate matrix.

macroecology: It refers to the study of statistical distributions of large groups of organisms or species to see why the distribution patterns vary.

macromolecule: It is a very large molecule most often created by some form of polymerization. In the context of biochemistry, the term may be applied to four conventional biopolymers (nucleic acids, proteins, carbohydrates, and lipids), as well as non-polymeric molecules with large molecular mass such as macrocycles.

macronutrient: In general, the macronutrients of plants are those elements which are required in larger quantities and they are carbon, hydrogen, oxygen, nitrogen, potassium, calcium, phosphorus, magnesium, and sulphur. In plant tissue culture, the term refers to the nutrients that are used in the growth media in concentrations >0.5 millimole/L. They are provided in the form of compounds generally called major salts.

macrophages: They are white blood cells present within the tissues, produced by the division of monocytes. Human macrophages are about 21 micrometres (0.00083 in) in diameter. Monocytes and macrophages are phagocytes, acting in both non-specific defence (innate immunity) as well as to help initiate specific defence mechanisms (adaptive immunity) of vertebrate animals.

macrophyte: It refers to a plant too large to be considered microscopic.

mad cow disease: It refers to the bovine spongiform encephalopathy (BSE), a fatal, infectious disease that degenerates the central nervous system of the cattle. It might have evolved from scrapie, a similar disease that infects sheep and goats.

major histocompatibility complex (MHC): It is a large genomic region or gene family found in most vertebrates. It is the most gene-dense region of the mammalian genome and plays an important role in the immune system and autoimmunity. The diversity of MHC is important in the immune diversity in population.

malignancy: It is the tendency of tumours to become progressively worse and potentially results in death. It is characterized by anaplasia, invasiveness, and metastasis.

malignant: It is a corresponding adjectival medical term used to describe a severe and progressively worsening disease. The term is most familiar as a description of cancer. A *malignant tumour* may be contrasted with a non-cancerous *benign tumour* in that a *malignancy* is not self-limited in its growth, is capable of invading into adjacent tissues, and may be capable of spreading to distant tissues (*metastasizing*), while a *benign tumour* has none of those properties. Malignant tumour is synonymous with cancer.

malt extract: It is a sticky, sugary substance obtained from malt soaked in water. It is used as a medicinal food.

malting: It is a process applied to cereal grains, in which the grains are made to germinate by soaking in water and then quickly halted from germinating further by drying/heating with hot air. Thus, malting is a combination of two processes: the sprouting process and the kiln-drying process.

mammary glands: They are the organs that, in mammals, produce milk for the sustenance of the young. These exocrine glands are enlarged and modified sweat glands and give mammals their name. The mammary glands of domestic mammals that have more than two breasts are called dugs.

mammary tumour: It is a tumour originating in the mammary gland. It is a common finding in older female dogs and cats that are not spayed, but they are found in other animals as well. The mammary glands in dogs and cats are associated with their nipples and extend from the underside of the chest to the groin on both sides of the midline.

mangrove: It refers to the trees that grow in coastal wetlands in tropical climates. Their trunks block wind, and their roots provide habitats and a barrier to erosion.

mannitol: It is an organic compound with the formula ($C_6H_8(OH)_6$). This polyol is used as an osmotic diuretic agent and a weak renal vasodilator. It was originally isolated from the secretions of flowering ash, called manna, after their resemblance to the Biblical food, and is also referred to as **mannite** and **manna sugar**.

mannose: It is a sugar monomer of aldohexose series of carbohydrates.

map distance: The standard measure of distance between loci, expressed in centiMorgans (cM). Estimated from recombination fraction via a mapping function. For small recombination fractions, map distance equals the percentage of recombination (recombination frequency) betwen two genes. 1% recombination = 1 cM. Sometimes called a map unit.

map unit: A unit that represents a recombination frequency of one per cent between genes and is used as a measure of distance between genes in the construction of genetic maps.

mapping (genome): It is the creation of a genetic map assigning DNA fragments to chromosomes. The map improves with the scientific progress and is perfect when the genomic DNA sequencing of the species has been completed. During this process, and for the investigation of differences in strain, the fragments are identified by small tags.

marginal probability: The unconditional probability of one event occurring; the probability of a single event.

marginal totals: The row and column totals of a contingency table are called marginal totals.

mariculture: It refers to growing and harvesting marine life.

marker peptide: A portion of fusion protein that facilitates its identification or purification.

marker: It is a gene or DNA sequence with a known location on a chromosome and associated with a particular gene or trait. It can be described as a variation, which may arise due to mutation or alteration in the genomic loci that can be observed. A genetic marker may be a short DNA sequence, such as a sequence surrounding a single base-pair change (single nucleotide polymorphism, SNP), or a long one, like minisatellites.

marker-assisted introgression: The use of DNA markers to increase the speed and efficiency of introgression of a new gene or genes into a population. The markers will be closely linked to the gene(s) in question.

marker-assisted selection (MAS): It is a process whereby a marker (morphological, biochemical or one based on DNA/RNA variation) is used for indirect selection of a genetic determinant or determinants of a trait of interest (i.e., productivity, disease resistance, abiotic stress tolerance, and/or quality). This process is used in plant and animal breeding.

Markov chain: It refers to any multivariate probability density whose independence diagram is a chain. The variables are ordered, and each variable "depends" only on its neighbours in the sense of being conditionally independent of the others. Markov chains are an integral component of hidden Markov models.

mass selection: Method of collection of seeds from desirable appearing individuals in a population, and the next generation is sown from the stock of mixed seed. This procedure, sometimes referred to as phenotypic selection, is based on how each individual looks. Mass selection has been used widely to improve old "land" varieties, varieties that have been passed down from one generation of farmers to the next.

maternal effect: It is a situation where the phenotype of an organism is determined not only by the environment it experiences and its genotype, but also by the environment and phenotype of its mother.

maternal inheritance: It is a type of uniparental inheritance in which all the progeny of a mating inherit the genotype and phenotype of the female parent.

mean: A central tendency measure representing the arithmetic average of a set of observations.

measure of central tendency: A measure indicating the value to be expected of a typical or middle data point.

measures of dispersion: A measure describing how scattered or spread out the observations in a data set are, etc.

median class: The class in a frequency distribution that contains the median value for a data set.

median: The middle point of a data set, a measure of location that divides the data set into halves.

medium (nutrient) or growth medium or culture medium: It is a liquid or solid medium designed to support the growth of microorganisms, cells, protoplasts, tissue, organs, or small plantlets under in vitro conditions.

medium formulation: It is a particular formula for preparing the culture medium. It commonly contains macro-elements and micro-elements (high and low salt), some vitamins (B vitamins, inositol), plant growth regulators (auxin, cytokinin and sometimes gibberellin), a carbohydrate source (usually sucrose or glucose) and often other substances, such as amino acids or complex growth factors. Media may be liquid or solidified with agar; the pH is adjusted and the solution is sterilized usually by filtration or autoclaving.

megabase (Mb): Unit of length for DNA fragments that is equal to 1 million nucleotides. A megabase (Mb) is roughly equal to 1 centiMorgan (cM).

megabase cloning: The cloning of very large DNA fragments.

megadalton (MDa): A unit equal to one million daltons.

megaspore (macrospore): A haploid (n) spore developing into a female gametophyte in heterosporous plants.

meiosis: It is a process of reductional division in which the number of chromosomes per cell is cut in half. In animals, meiosis always results in the formation of gametes, while in other organisms it can give rise to spores. As with mitosis, before meiosis begins, the DNA in the original cell is replicated during S-phase of the cell cycle. Two cell divisions separate the replicated chromosomes into four haploid gametes or spores.

meiotic analysis: A technique used to analyse chromosome-pairing relationships.

meiotic drive: Preferential selection during meiosis (germ cell production). In meiotic drive, there is a preferential production of certain gametes. This alters the segregation of genes from the Mendelian expectations. Meiotic drive is a mechanism for transmission distortion.

melanin: It is a class of compounds found in plants, animals, and protists, where it serves predominantly as a pigment. The class of pigments are derivatives of the amino acid tyrosine. Many melanins are insoluble salts and show affinity to water.

melting (of DNA): It refers to the denaturation of double-stranded DNA into two single strands by the application of heat. (Denaturation breaks the hydrogen bonds holding the double-stranded DNA together.)

melting temperature (Tm): It is the process by which double-stranded deoxyribonucleic acid unwinds and separates into single-stranded strands through the breaking of hydrogen bonding between the bases.

membrane bioreactors: It is the combination of a membrane process like microfiltration or ultrafiltration with a suspended growth bioreactor, and is now widely used for municipal and industrial wastewater treatment with plant sizes up to 80,000 population equivalent.

memory cells: The term is usually used with respect to immunity. In the wake of the first (primary response) infection involving a particular antigen, the responding naïve (ones which have never been exposed to the antigen) cells proliferate to produce a colony of cells, most of which differentiate into the plasma cells, also called effector B cells (which produce the antibodies) and clear away with the resolution of infection, and the rest persist as the memory cells that can survive for years, or even a lifetime.

memory: It is an organism's ability to store, retain, and recall information.

Mendel's Laws: The principles of heredity were written by Gregor Mendel, who summarized his findings in two laws, the **Law of Segregation** and the **Law of Independent Assortment**.

Mendelism: It is a set of primary tenets relating to the transmission of hereditary characteristics from parent organisms to their offspring. They were initially derived from the work of Gregor Mendel published in 1866 and became the core of classical genetics.

mericlinal: Refers to a chimera with tissue of one genotype partly surrounded by that of another genotype.

mericloning: It is the vegetative culture of a plant through the tissue culture of apical meristems. Plants produced through this process are relatively identical.

meristem culture: Culture of meristem tips was started first by Morel and co-workers in the 1950's with the intention of obtaining virus-free plants from those already infected.

meristem tip: An explant comprising the meristem (meristematic dome) and usually one pair of leaf primordia.

meristemoid: A small, triangular stomatal precursor cell that functions temporarily as a stem cell in a meristem.

merozygote: It is a state when a cell, usually bacterial, is temporarily diploid as result of DNA transfer processes like conjugation.

mesokurtic: A moderately peaked distribution.

mesophile: It is an organism that grows best in moderate temperature, neither too hot nor too cold, typically between 15 and 40°C.

messenger RNA (mRNA): It is a molecule of RNA encoding a chemical "blueprint" for a protein product. mRNA is transcribed from a DNA template, and carries coding information to the sites of protein synthesis: the ribosomes. Here, the nucleic acid polymer is translated into a polymer of amino acids: a protein.

metabolic cell: A cell that is not dividing.

metabolism: It is a set of chemical reactions that occur in living organisms to maintain life. These processes allow organisms to grow and reproduce, maintain their structures, and respond to their environments. Metabolism is usually divided into two categories—catabolism, breaks down organic matter, for example, to harvest energy in cellular respiration; anabolism, uses energy to construct components of cells such as proteins and nucleic acids.

metabolites: They are the intermediates and products of metabolism. The term *metabolite* is usually restricted to small molecules.

metacentric: A chromosome is metacentric if its two arms are roughly equal in length.

metaheuristics: It is a computational method that optimizes a problem by iteratively trying to improve a candidate solution with regard to a given measure of quality. Metaheuristics make few or no assumptions about the problem being optimized and can search very large spaces of candidate solutions. However, metaheuristics do not guarantee an optimal solution.

metallothionein: It is a family of cysteine-rich, low molecular weight (MW ranging from 3500 to 14000 Da) proteins.

metamorphism: It refers to the transformation of one to the other.

metaphase: It is a stage of mitosis in the eukaryotic cell cycle in which condensed and highly coiled chromosomes, carrying genetic information, align in the middle of the cell before being separated into each of the two daughter cells. Metaphase accounts for approximately 4% of the cell cycle's duration.

metapopulation: It refers to local but geographically separated populations of a species that are linked only by organisms that migrate between them. Although local groups might die out, the migrants keep the species alive.

metastasis: It is the spread of a disease from one organ or part to another non-adjacent organ or part.

methionine: It is an α-amino acid with the chemical formula $HO_2CCH(NH_2)CH_2CH_2SCH_3$. It is abbreviated as Met or M. It is an essential amino acid and is classified as nonpolar.

methylation: The addition of a methyl group (–CH3) to a macromolecule, such as the addition of a methyl group to specific cytosine and, occasionally, adenine residues in DNA.

Michaelis constant: It approximately describes the kinetics of many enzymes and is mathematically identical to Langmuir-Hinshelwood kinetics. It is named after Leonor Michaelis and Maud Menten. The Michaelis-Menten equation relates the initial reaction rate v_0 to the substrate concentration [S].

microalgal culture: Culture in bioreactors of microalgae; microalgae include seaweeds.

microarray: It is a multiplex lab-on-a-chip. It is a 2D array on a solid substrate (usually a glass slide or silicon thin-film cell) that assays large amounts of biological material using high-throughput screening methods.

microbe: A microscopic living organism, such as a bacterium, fungus, protozoan or virus.

microbes: They are thought to make up more than 60 per cent of the earth's biomass. They have survived and evolved on the earth for over 3.7 billion years and have been found in almost every environment. The diversity and range of environmental adaptations mean that microbes have long ago "solved" many problems for which scientists are still seeking solutions. In addition, most microbes are not responsible for diseases in humans, animals, or plants. Therefore, by studying and understanding a diverse group of microbes at the genomic level, researchers may develop answers to the existing challenges in medicine, agriculture, industrial processes, energy production and use, and environmental and waste clean-up.

microbial mat: It is a multilayered sheet of microorganisms, mainly bacteria and archaea.

microbody: It is a cytoplasmic organelle of a more or less globular shape that comprises degradative enzymes bound within a single membrane. Microbodies are specialized as containers for metabolic activity. Types include peroxisomes, glyoxisomes, glycosomes and Woronin bodies.

microcarriers: They are small particles (0.2 mm) used for the growth of anchorage-dependent cells. They were described by Van Wezel in 1967. These microcarriers are suspended in the culture medium by gentle agitation and a homogeneous environment is obtained. The scale-up is done by increasing the fermentor volume.

microclimate: It refers to the climate of small locales: under plant leaves, in a garden, on a hillside.

microdroplet array (multiple drop array (MDA) or hanging droplet technique: The technique was introduced by Kao and Konstabel. It is used to evaluate large numbers of media modifications, employing small quantities of medium into which small numbers of cells are placed for investigation. Static microdroplet array developed by Heubner et al., in 2008 is a single-layered poly (dimethylsiloxane) microfluidic structure for the entrapment and release of microdroplets in an array format controlled entirely by liquid flow. Aqueous picoliter droplets are trapped en masse and optically monitored for extended periods of time. Such an array-based approach is used to characterize droplet shrinkage, aggregation of encapsulated *E. coli* cells and enzymatic reactions.

microelement: A trace element required in very small quantities.

microencapsulation: It is a process in which tiny particles or droplets are surrounded by a coating to give small capsules many useful properties. In a relatively simplistic form, a microcapsule is a small sphere with a uniform wall around it.

microenvironment: A climate change link library with comprehensive global solutions.

microfibrils: It is a very fine fibril, or fibre-like strand, consisting of glycoproteins and cellulose. It is usually, but not always, used as a general term while describing the structure of protein fibre, for example, hair and sperm tail.

microfluidics: It refers to the miniaturization of chemical reactions or pharmacological assays into microscopic tubes or vessels in order to greatly increase their throughput, by placing many of them side-by-side in an array.

microinjection: It refers to the process of using a glass micropipette to insert substances at a microscopic or borderline macroscopic level into a single living cell. It is a simple mechanical process in which a needle, roughly 0.5 to 5 micrometre in diameter, penetrates the cell membrane and/or the nuclear envelope. The desired contents are then injected into the desired subcellular compartment and the needle is removed.

microisolating system: It is a process of mechanical separation of single cells or protoplasts, thus allowing them to proliferate individually.

micron (micrometre): It is one millionth of a metre, or equivalently one thousandth of a millimetre or one thousand nanometres. It can also be written in scientific notation as 1×10^{-6} m, meaning $^{1}/_{1000000}$ m.

micronuclei assay: This assay is used when a large number of people are exposed to a disastrous situation, for instance, when an atom bomb is dropped, as was done at Hiroshima and Nagasaki in 1945. This assay is used as a screening procedure when a large number of people happen to be exposed, as scoring of micronuclei under the microscope is much easier compared to scoring dicentric chromosomes. This assay is however comparatively less sensitive than dicentric measurement.

micronutrients: These are the nutrients needed throughout life in small quantities. They are dietary minerals needed by the human body in very small quantities as opposed to macrominerals which are required in larger quantities. The microminerals or trace elements include at least iron, cobalt, chromium, copper, iodine, manganese, selenium, zinc and molybdenum.

microorganism: Very small organisms visible only under a microscope.

microplasts: Vesicles produced by subdivision and fragmentation of protoplasts or thin-walled cells.

micropropagation: It is the practice of rapidly multiplying stock plant material to produce a large number of progeny plants, using modern plant tissue culture methods. Micropropagation is used to multiply novel plants, such as those that have been genetically modified or bred through conventional plant breeding methods. It is also used to provide a sufficient number of plantlets for planting from a stock plant which does not produce seeds, or does not respond well to vegetative reproduction.

micropyle: A small opening in the surface of a plant ovule through which the pollen tube passes prior to fertilization.

microsatellite: These are simple sequence repeats (SSRs), or tandem repeats, repeating sequences of 1-6 base pairs of DNA. Microsatellites are typically neutral and co-dominant. They are used as molecular markers in genetics, for kinship, population and other studies.

microtuber: Cultured tissue capable of growing into tuberous plant.

microtubules: They are one of the components of cytoskeleton. They have a diameter of 25 nm and length varying from 200 nanometres to 25 micrometres. Microtubules serve as structural components within the cells and are involved in many cellular processes including mitosis, cytokinesis, and vesicular transport.

middle lamella: It is a pectin layer which cements the cell walls of two adjoining cells together. Plants need this to give them stability and so that they can form plasmodesmata between the cells. It is the first layer which is deposited at the time of cytokinesis. The middle lamella is made up of calcium and magnesium pectate. In plants, the pectins form a unified and continuous layer between adjacent cells.

mid-parent value: It is the value defined as an average of the trait value of father and mother.

mimetics: It refers to the compounds that mimic the function of other molecules via their high degree of structural (conformational) similarity, and hence physio-chemical properties.

minimum effective cell density: The inoculum density below which the culture fails to give reproducible cell growth. The minimum density is a function of the tissue (species, explant, cell line) and the culture phase of the inoculum suspension. Minimum density decreases inversely to the aggregate size and division rate of the stock culture.

minimum inoculum size: It is the critical volume of inoculum required to initiate culture growth due to diffusive loss of cell materials into the medium.

mini-prep: It refers to the minipreparation of plasmid DNA. It is a rapid, small-scale isolation of plasmid DNA from bacteria. It is based on the alkaline lysis method invented by the researchers Birnboim and Doly in 1979.

minor histocompatibility antigen: They are receptors on the cell surface of donated organs that are known to give an immunological response in some organ transplants. They cause problems of rejection less frequently than those of the major histocompatibility complex (MHC). Minor histocompatibility antigens are due to normal proteins that are in themselves polymorphic in a given population. Even when a transplant donor and recipient are identical with respect to their major histocompatibility complex genes, the amino acid differences in minor proteins can cause the grafted tissue to be slowly rejected.

mismatch repair (MMR): It is strand-specific repair mechanism. During DNA synthesis the newly synthesised (daughter) strand will include errors commonly. In order to correct this the mismatch repair machinery distinguishes the newly synthesised strand from the template (parental) strand. It is an evolutionarily conserved process that corrects mismatches generated during DNA replication and escape proofreading. MMR proteins also participate in many other DNA transactions, such that inactivation of MMR can have wide-ranging biological consequences, which can be either beneficial or detrimental.

missense mutation: It is a point mutation in which a single nucleotide is changed, resulting in a codon that codes for a different amino acid (mutations that change an amino acid to a stop codon are considered nonsense mutations, rather than missense mutations).

missense mutation: It refers to the point mutation in which one codon (triplet of bases) is changed into another designating a different amino acid.

mist propagation: It is the application of fine droplets of water to leafy cuttings in the rooting stage to reduce transpiration.

mitochondrial DNA (mtDNA): It is the DNA located in organelles called mitochondria, structures within the cells that convert the energy from food into a form that cells can use. Most other DNA present in eukaryotic organisms is found in the cell nucleus.

mitochondrion: It is a membrane-enclosed organelle found in most eukaryotic cells.

mitosis: It is the process by which a eukaryotic cell separates the chromosomes in its cell nucleus into two identical sets in two nuclei. Errors in mitosis can either kill a cell through apoptosis or cause mutations that may lead to cancer.

mixoploid: Having cells with different ploidy (different numbers of chromosomes).

mixotroph: It refers to an organism that can double as an autotroph (produces its own food) or heterotroph (eats food produced by other organisms).

mobilization: The transfer between bacteria of a non-conjugative plasmid by a conjugative plasmid.

mobilizing functions: It refers to the ability of genes on a plasmid that give it the aptitude to facilitate the transfer of either a non-conjugative or a conjugative plasmid from one bacterium to another.

modal class: In a frequency distribution, the class having the greatest frequency.

mode: The value more often repeated in the data set. It is represented by the highest pointing the distribution curve of a data set.

model organism: It is a non-human species that is extensively studied to understand particular biological phenomena, with the expectation that discoveries made in the organism model will provide insight into the workings of other organisms. In particular, model organisms are widely used to explore potential causes and treatments for human diseases when human experimentation would be unfeasible or unethical.

model theory: It is the study of mathematical structures such as groups, fields, graphs, or even universes of set theory, using tools from mathematical logic.

modelling: In bioinformatics, modelling usually refers to molecular modelling, a process whereby the three-dimensional architecture of biological molecules is interpreted (or predicted), visually represented, and manipulated in order to determine their molecular properties (general). A series of mathematical equations or procedures which simulate a real-life process, given a set of assumptions, boundary parameters, and initial conditions.

modifying gene: It is a gene that alters or influences the expression function of another gene, including the suppression or reduction of the usual function of the modified gene.

MOET(multiple ovulation and embryo transfer): It is the process whereby superior donor ewes are superovulated and artificially inseminated to produce fertilised embryos. These embryos are surgically removed at days 5-6 and then laparoscopically transferred to synchronised, commercial grade recipient ewes to establish a surrogate pregnancy.

molality (mol/kg, molal, or *m*): It denotes the number of moles of *solute* per kilogram of *solvent* (not *solution*).

molarity (mol/L, molar, or M or molar concentration): It denotes the number of moles of a given substance per litre of solution. A capital letter M is used to abbreviate units of mol/L.

mole (mol): It is the SI base unit of amount of substance; one of a few units used to measure this physical quantity.

mole cancer: It is the name given to the most dangerous form of skin cancer. The melanoma is so dangerous that once it grows to a certain thickness, it metastasizes or spreads throughout the body. After melanoma has spread to the internal organs, there is little that can be done and death follows shortly after.

mole fraction: It is a way of expressing the composition of a mixture. The mole fraction of each component *i* is defined as its amount of substance n_i divided by the total amount of substance in the system, *n*.

molecular biology: It is the study of biology at a molecular level. The field overlaps with other areas of biology and chemistry, particularly genetics and biochemistry. Molecular biology chiefly concerns itself with understanding the interactions between the various systems of a cell, including the interactions between DNA, RNA and protein biosynthesis as well as learning how these interactions are regulated.

molecular cloning: It refers to the procedure of isolating a defined DNA sequence and obtaining multiple copies of it in vivo. Cloning is frequently employed to amplify DNA fragments containing genes, but it can be used to amplify any DNA sequence such as promoters, non-coding sequences, chemically synthesised oligonucleotides and randomly fragmented DNA. Cloning is used in a wide array of biological experiments and technological applications such as large scale protein production.

molecular genetics: The area of knowledge concerned with the genetic aspects of molecular biology, especially DNA, RNA and protein molecules.

molecular machine (nanomachine): It is defined as a discrete number of molecular components which perform mechanical-like movements (output) in response to specific stimuli (input). More generally, the expression is often applied to molecules that simply mimic functions which occur at the macroscopic level. The term is also common in nanotechnology, and a number of highly complex molecular machines have been proposed towards the goal of constructing a molecular assembler. Molecular machines can be divided into two broad categories: synthetic and biological.

molecular modelling: It is a collective term that refers to theoretical methods and computational techniques to model or mimic the behaviour of molecules. The techniques are used in the fields of computational chemistry, computational biology and materials science for studying molecular systems ranging from small chemical systems to large biological molecules and material assemblies. The simplest calculations can be performed by hand, but inevitably computers are required to perform molecular modelling of any reasonably sized system. The common feature of molecular modelling techniques is the atomistic level description of the molecular systems; the lowest level of information is individual atoms (or a small group of atoms).

molecule: It is defined as an electrically neutral group of at least two atoms in a definite arrangement held together by very strong (covalent) chemical bonds. Molecules are distinguished from polyatomic ions in this strict sense. In organic chemistry and biochemistry, the term *molecule* is used less strictly and is also applied to charged organic molecules and biomolecules.

monoculture: It is the agricultural practice of producing or growing one single crop over a wide area. It is usually developed by extensive growing farmers. The technique is widely used in modern industrial agriculture and its implementation has allowed for large harvests from minimal labour. However, monoculture have weaknesses, such as an increased susceptibility to disease that can make them less suitable than polycultures for growing certain crops.

monoclonal antibody (mAb or moAb): They are monospecific antibodies that are the same because they are made by one type of immune cell which are all clones of a unique parent cell. Given almost any substance, it is possible to create monoclonal antibodies that specifically bind to that substance; they can then serve to detect or purify that substance.

monoculture: It refers to the cultivation of a single crop on a piece of land to the exclusion of other crops. This generally requires large quantities of artificial fertilizers, herbicides, pesticides, nematocides, and other potential toxic measures to kill bugs and increase the yield.

monoecious: It denotes the plant species that have separate male and female flowers on the same plant.

monogastric animals: Animals with simple stomachs that do not ruminate.

monogenic disorder: It results from a single mutated gene. There are estimated to be over 4000 human diseases caused by single gene defects. Single gene disorders can be passed on to subsequent generations in several ways.

monohybrid cross: It is a cross between parents who are heterozygous at one locus. Monohybrid inheritance is the inheritance of a single characteristic. The different forms of characteristics are usually controlled by different alleles of the same gene.

monolayer: It is a one-molecule thick layer of an insoluble organic material spread onto an aqueous subphase. Traditional compounds used to prepare Langmuir monolayers are amphiphilic materials that possess a hydrophilic headgroup and a hydrophobic tail.

monophagus: It refers to predators that favour a single prey species.

monophyletic group: It is a group of organisms which forms a clade, meaning that it consists of an ancestor and all its descendants.

monosaccharides: They are the most basic units of biologically important carbohydrates. They are the simplest form of sugar and are usually colourless, water-soluble, crystalline solids.

monosomic genome: A genome that is basically diploid but has only one copy of one particular chromosome type so that its chromosome number is 2n-1. An XO human individual is monosomic for the X-chromosome.

monozygotic twins: They are the identical twins. They occur when a single egg is fertilized to form one zygote (monozygotic) which then divides into two separate embryos.

morphogen: It is a hypothetical substance governing the pattern of tissue development and, in particular, the positions of the various specialized cell types within a tissue. It spreads from a localized source and forms a concentration gradient across a developing tissue.

morphogenesis: It is the biological process that causes an organism to develop its shape. It is one of the three fundamental aspects of developmental biology along with the control of cell growth and cellular differentiation.

morphogenic response: It refers to the effect on the developmental history of a plant or its parts exposed to a given set of growth conditions, or to a change in the environment.

morphology: It is the study of form, structure and configuration of an organism.

mosaic: An organism or part of an organism that is composed of cells with different origin.

mother plant: It is a gardening term that describes a plant from which you are going to take cuttings.

motif: It refers to the conserved element of a protein sequence alignment that usually correlates with a particular function. Motifs are generated from a local multiple protein sequence alignment corresponding to a region whose function or structure is known. It is sufficient that it is conserved, and is likely to be predictive of any subsequent occurrence of such a structural/functional region in any other novel protein sequence.

Mouse (*Acomys cahirinus*): It is the mammalian model organism closest to humans. The gene sequences that code for numerous proteins responsible for carrying out vital biological processes in both the human and the mouse share a high degree of similarity. Therefore, the mouse has already proven extremely useful in development, genetic, and immunology studies. Elucidation of the mouse genomic sequence will also provide a system for studying and understanding human disease, as well as a mechanism for investigating new treatment strategies in ways that cannot be done in humans.

movable genetic element (transposon): These are sequences of DNA that can move around to different positions within the genome of a single cell, a process called transposition.

mRNA: See messenger RNA.

multigene family: It refers to the set of genes derived by duplication of an ancestral gene, followed by independent mutational events resulting in a series of independent genes either clustered together on a chromosome or dispersed throughout the genome.

multigene family: This term is used to include groups of genes from the same organism that encodes proteins with similar sequences either over their full lengths or limited to a specific domain.

multigenic: Controlled by several genes, as opposed to monogenic.

multilocus probe: A probe that hybridizes to a number of different sites in the genome of an organism.

multimer: A protein made up of more than one peptide or a chain of organic polymer consisting of two or more organic molecules; dimethacrylate (Bis-GMA) and urethane dimethacrylate (UDMA) oligomers are the basis of most common resins.

multiple (sequence) alignment: It refers to the multiple alignment of *k* sequences in a rectangular array, consisting of characters taken from the alphabet A, that satisfies the following conditions: There are exactly *k* rows; ignoring the gap character, row number *i* is exactly the sequence s_i; and each column contains at least one character different from "—". In practice multiple sequence alignments include a cost/weight function, that defines the penalty for the insertion of gaps (the "—" character) and weights identities and conservative substitutions accordingly. Multiple alignment algorithms attempt to create the optimal alignment defined as the one with the lowest cost/weight score.

multiple alleles: Any of a set of three or more alleles, or alternative states of a gene, only two of which can be present in a diploid organism.

multiple bar diagram: The technique of simple bar diagrams can be extended to represent two or more sets of interrelated data in a diagram.

multiple cloning site (MCS) or polylinker: It is a short segment of DNA which contains many (up to ~20) restriction sites—a standard feature of engineered plasmids. Restriction sites within an MCS are typically unique, occurring only once within a given plasmid. MCSs are commonly used during the procedures involving molecular cloning or subcloning.

multiplex sequencing: It refers to the approach to high-throughput sequencing that uses several pooled DNA samples run through gels simultaneously and then separated and analyzed.

multivalent vaccine (polyvalent vaccine): It is the one prepared from cultures or antigens of more than one strain or species.

mutable genes: Any of a class of unstable genes that spontaneously mutate at a sufficiently high rate to produce mosaicism.

mutagen: It is a physical or chemical agent that changes the genetic material, usually DNA, of an organism and thus increases the frequency of mutations above the natural background level.

mutagenesis: It is a process by which the genetic information of an organism is changed in a stable manner, either in nature or experimentally by the use of chemicals or radiation.

mutant: It is an individual organism, or a new genetic character arising or resulting from an instance of mutation, which is a base-pair sequence change within the DNA of a gene or chromosome of an organism resulting in the creation of a new character or trait not found in the wild type.

mutation pressure: The constant resupplying of mutations to a gene pool due to a base mutation rate.

mutations: These are changes in the DNA sequence of a cell's genome and are caused by radiation, viruses, transposons and mutagenic chemicals, as well as errors that occur during meiosis or DNA replication.

mutualism: It refers to a relationship between the species that benefits both.

mutually exclusive events: Events that cannot happen together.

mycelium: It is the vegetative part of a fungus, consisting of a mass of branching, thread-like hyphae.

mycorrhiza: It is a symbiotic (generally mutualistic, but occasionally weakly pathogenic) association between a fungus and the roots of a plant.

mycotoxin: It is a toxic secondary metabolite produced by the fungus. It is usually reserved for the toxic chemical products produced by fungi that readily colonize crops. Most fungi are aerobic and are found almost everywhere in extremely small quantities due to the minute size of their spores. They consume organic matter wherever humidity and temperature are sufficient. One mold species may produce many different mycotoxins and/or same mycotoxin as another species.

myeloma: It is a cancer of white blood cells known as plasma cells. A type of B cell, plasma cells are a crucial part of immune system responsible for the production of antibodies in humans and other vertebrates. They are produced in the bone marrow and are transported through the lymphatic system.

myo inositol (inositol or **cyclohexane-1, 2, 3, 4, 5, 6-hexol):** It is a sixfold alcohol (polyol) of cyclohexane. It exists in nine possible stereoisomers. It plays an important role as the structural basis for a number of secondary messengers in eukaryotic cells, including inositol phosphates, phosphatidylinositol (PI) and phosphatidylinositol phosphate (PIP) lipids. In plants, the hexaphosphate of inositol is found as phytic acid.

myosin: They are a large family of motor proteins found in eukaryotic tissues.

N

N_2: It is the chemical formula for nitrogen. Nitrogen is a chemical element that has the symbol **N**.

nanobiotechnology: It is the branch of nanotechnology with biological and biochemical applications or uses. Nanobiotechnology often studies existing elements of nature in order to fabricate new devices.

nanoengineering: It is the practice of engineering on the nanoscale. It derives its name from the nanometre, a unit of measurement equalling one billionth of a metre.

nanofluidics: It is the study of behaviour, manipulation, and control of fluids that are confined to structures of nanometre (typically 1-100 nm) characteristic dimensions (1 nm = 10^{-9} m). Fluids confined in these structures exhibit physical behaviours not observed in larger structures, such as those of micrometre dimensions and above, because the characteristic physical scaling lengths of the fluid, very closely coincide with the dimensions of the nanostructure itself.

nanometre: It is a unit of length in the metric system, equal to one billionth of a metre. It is one of the more often used units for very small lengths, and equals ten Ångström, an internationally recognised non-SI unit of length.

nanoparticles: It is a small object that behaves as a whole unit in terms of its transport and properties. It is further classified according to size: in terms of diameter, fine particles cover a range between 100 and 2500 nanometers, while ultrafine particles, on the other hand, are sized between 1 and 100 nanometres.

nanorobotics: It is the technology of creating machines or robots at or close to the microscopic scale of a nanometer (10^{-9} metres). More specifically, nanorobotics refers to the still largely hypothetical nanotechnology engineering discipline of designing and building nanorobots, devices ranging in size from 0.1-10 micrometres and constructed of nanoscale or molecular components. Nobody has yet built artificial non-biological nanorobots: they remain a hypothetical concept. The names nanobots, nanoids, nanites or nanomites have also been used to describe these hypothetical devices.

nanotechnology: It is the study of controlling of matter on an atomic and molecular scale. Generally, nanotechnology deals with the structures of the size 100 nanometres or smaller in at least one dimension, and involves developing materials or devices within that size. Nanotechnology is very diverse, ranging from extensions of conventional device physics to completely new approaches based upon molecular self-assembly, from developing new materials with dimensions on the nanoscale to investigating whether we can directly control matter on the atomic scale.

narrow-host-range plasmid: A plasmid that can replicate in one, or at most a few, different bacterial species.

narrow-sense heritability: In quantitative genetics, the proportion of the phenotypic variance that is due to variation in breeding values.

native protein: The concept of a protein in its natural state, in the cell, unaltered by heat, chemicals, enzyme action, or the exigencies of extraction.

natural selection: It is the process whereby heritable traits that make it more likely for an organism to survive and successfully reproduce become more common in a population over successive generations. It is a key mechanism of evolution.

NCEs (New Chemical Entity): It refers to the compounds identified as potential drugs that are sent from research and development into clinical trials to determine their suitability.

necrosis: It is the premature death of cells and living tissue. Necrosis is caused by external factors, such as infection, toxins or trauma. This is in contrast to apoptosis, which is a naturally occurring cause of cellular death. While apoptosis often provides beneficial effects to the organism, necrosis is almost always detrimental, and can be fatal.

negative autogenous regulation (negative self-regulation): Inhibition of the expression of a gene or set of coordinately regulated genes by the product of the gene or the product of one of the genes.

negative control system: It is a mechanism in which the regulatory protein is required to turn off the gene expression.

negative correlation: The values of two variables move in the reverse direction.

negative feedback system: It occurs when the output of a system acts to oppose changes to the input of the system, with the result that the changes are attenuated. If the overall feedback of the system is negative, then the system will tend to be stable.

negative selection: It is a process in natural selection, in which there is a selective removal of alleles that are deleterious. This can result in stabilizing selection through the purging of deleterious variations that arise. It is also known as purifying selection.

nematodes: It refers to microscopic cylindrical roundworms living in water and soil.

neoformation: It is a new and abnormal growth of tissue, tumour or neoplasm. It

neoteny (juvenilization): It is the retention, by adults in a species, of traits previously seen only in juveniles.

nested PCR: It refers to the second round amplification of an already PCR-amplified sequence using a new pair of primers which are internal to the original primers. Typically done when a single PCR reaction generates insufficient amounts of product.

neural net: Neural net is an interconnected assembly of simple processing elements, units or nodes, whose functionality is loosely based on the animal brain. The processing ability of the network is stored in the inter-unit connection strengths, or weights, obtained by a process of adaptation to, or learning from, a set of training patterns.

neuron: It refers to a nerve cell that transmits electrochemical impulses. When triggered at the dendrite end by an incoming nerve pulse, they generate a pulse down the axon to the terminal button, where the pulse triggers the release of neurotransmitter chemicals into the synpase, the space between one neuron and another. These chemicals then trigger a pulse (or inhibit one) in other neurons.

neurotoxin: It refers to a poison that disrupts nerve function. Some venoms are neurotoxins that paralyze the prey.

neutral mutation: It is a mutation that occurs in an amino acid codon (presumably within an mRNA molecule), which results in the use of a different (but often chemically similar) amino acid that has a negligible effect on fitness.

neutral theory (neutral theory of molecular evolution): It is a theory proposed by Motoo Kimura, which states that the vast majority of evolutionary changes at the molecular level are caused by random drift of selectively neutral mutants.

NFT (neurofibrillary tangles): They are pathological protein aggregates found within the neurons in cases of Alzheimer's disease.

nick translation: It is a tagging technique in molecular biology in which DNA Polymerase I is used to replace some of the nucleotides of a DNA sequence with their labelled analogues, creating a tagged DNA sequence, which can be used as a probe in fluorescent in situ hybridization or blotting techniques.

nick: It is a break in phosphodiester bond in the backbone of one of the strands of a duplex DNA molecule.

nicked circle (relaxed circle): During the extraction of plasmid DNA from the bacterial cell, one strand of the DNA becomes nicked. This relaxes the torsional strain needed to maintain supercoiling, producing the familiar form of plasmid.

nitrate: It is a salt of nitric acid with an ion composed of one nitrogen and three oxygen atoms (NO_3^-). In organic chemistry, the esters of nitric acid and various alcohols are called nitrates. Nitrogen trioxide is an alternative name for nitrate.

nitrification: It is the biological oxidation of ammonia into nitrite followed by the oxidation of nitrites into nitrates. Degradation of ammonia to nitrite is usually the rate limiting step of nitrification.

nitrite: The nitrite ion is NO_2^-. The anion is bent, being isoelectronic with ozone. More generally, a nitrite compound is either a salt or an ester of nitrous acid.

nitrocellulose (cellulose nitrate): It is a highly flammable compound formed by nitrating cellulose through exposure to nitric acid or another powerful nitrating agent.

nitrogen assimilation: It is a fundamental biological process that occurs in plants and algae that are incapable of independent nitrogen fixation. The assimilation of nitrogen has marked effect on plant productivity, biomass, and crop yield, and nitrogen deficiency leading to a decrease in structural components.

nitrogen fixation: It refers to the biological process by which nitrogen (N_2) in the atmosphere is converted into ammonia.

nitrogenous bases: These are the organic compounds that owe their property as a base to the lone pair of electrons of a nitrogen atom. In biological sciences, nitrogenous bases are typically classified as the derivatives of two parent compounds, pyrimidine and purine. Their flat shape is particularly important when considering their role in nucleic acids as nucleobases (building blocks of DNA and RNA): adenine, guanine, thymine, cytosine, and uracil. These nitrogenous bases hydrogen bond between opposing DNA strands to form the rungs of the "twisted ladder" or double mount helix of DNA.

nod box: A DNA sequence that controls the transcriptional regulation of *Rhizobium* nodulation genes.

nodal culture: The culture of a lateral bud and a section of adjacent stem tissue.

node: It refers to the point on a stem from where new stem or leaves grow.

nodular melanoma: It is the most aggressive form of melanoma. It grows in vertical direction from the outset and grows very fast (months). Nodular melanoma has no known precursor. It is a small black, or if amelanotic, pink nodule that simply enlarges. The lesions tend to bleed.

nodular: It refers to a pebbly (rough) texture of a callus produced usually in plant tissue culture.

nodulation: It is the formation of nodules by symbiotic bacteria on the roots of plants.

nodule (root): Root nodules occur on the roots of plants that associate with symbiotic nitrogen-fixing bacteria. Under nitrogen limiting conditions, plants from the pea family Fabaceae form a symbiotic relationship with a host-specific strain of bacteria known as rhizobia.

non-disjunction: It is the failure of chromosome pairs to separate properly during cell division. This could arise from a failure of homologous chromosomes to separate in meiosis-I, or the failure of sister chromatids to separate during meiosis-II or mitosis.

non-histone chromosomal proteins: Apart from histones, the chromatin consists of DNA polymerases, regulator proteins, etc. They are often lumped together called nonhistone proteins.

non-sampling error: Errors that mainly arise at the stages of observation and processing of the data are called non-sampling errors.

nonsense mutation: It is a point mutation in a sequence of DNA that results in a premature stop codon, or a nonsense codon in the transcribed mRNA, and in a truncated, incomplete, and usually nonfunctional protein product. It differs from missense mutation, which is a point mutation where a single nucleotide is changed to cause substitution of a different amino acid.

non-target organism: An organism which is affected by an interaction for which it was not the intended recipient.

non-template strand: It is the non-transcribed strand of DNA or sense strand or coding strand. It will have the same sequence as the RNA transcript, except that T is present at positions where U is present in the RNA transcript.

NOR (nucleolar organizer region): It is a chromosomal region around which the nucleolus forms. This region is the particular part of a chromosome that is associated with a nucleolus after the nucleus divides. The region contains several tandem copies of ribosomal RNA genes.

normal distribution: A distribution of a continuous random variable with a bell-shaped curve. The mean lies at the centre of the distribution and the curve is symmetrical. The two tails extended indefinitely, never touching the horizontal axis.

Northern blotting: It is a technique used in molecular biology research to study gene expression by detection of RNA (or isolated mRNA) in a sample. It is similar to Southern blotting, except that RNA is transferred onto a matrix and the presence of a specific RNA molecule is detected by DNA-RNA hybridization.

Northern hybridization: Hybridization of a labelled DNA probe to RNA fragments that have been transferred from an agarose gel to a nitrocellulose filter.

nucellar embryo: It is an embryo which has developed vegetatively from somatic tissue surrounding the embryo sac, rather than by fertilization of the egg cell.

nucellus: The central portion of an ovule in which the embryo sac develops.

nuclease protection assay: It is a laboratory technique used in biochemistry and genetics to identify individual RNA molecules in a heterogeneous RNA sample extracted from cells.

nuclease: It is an enzyme capable of cleaving the phosphodiester bonds between the nucleotide subunits of nucleic acids.

nucleic acid hybridization probe: It is a fragment of DNA or RNA of variable length (usually 100-1000 bases long), which is used in DNA or RNA samples to detect the presence of nucleotide sequences (the DNA target) that are complementary to the sequence in the probe. The probe thereby hybridizes to single-stranded nucleic acid (DNA or RNA) whose base sequence allows probe-target base pairing due to complementarity between the probe and target.

nucleic acid probe: It is a nucleic-acid fragment that is complementary to another nucleic-acid sequence and thus, when labelled in some manner, as with a radioisotope, can be used to identify complementary segments present in the nucleic acid sequences of various microorganisms.

nucleic acid: It is a macromolecule composed of chains of monomeric nucleotides. In biochemistry these molecules carry genetic information or form structures within cells. The most common nucleic acids are **deoxyribonucleic acid** (DNA) and **ribonucleic acid** (RNA).

nuclein: It refers to any of a group of decomposition substances found in the nuclei of cells that are intermediate to nucleoproteins and nucleic acids.

nucleo-cytoplasmic ratio: It is the ratio between the measured cross-sectional area or estimated volume of the nucleus of a cell to the volume of its cytoplasm. Also known as karyoplasmic ratio.

nucleolar organizer region (NOR): It is a chromosomal region around which the nucleolus forms. This region is the particular part of a chromosome that is associated with a nucleolus after the nucleus divides.

nucleolar zone: It refers to any chromosome region, irrespective of whether or not it is a secondary-constriction, that is associated with the formation of nucleolus during telophase.

nucleolus: It is a non-membrane bound structure composed of proteins and nucleic acids found within the nucleus.

nucleoplasm: It is a highly viscous liquid that surrounds the chromosomes and nucleoli.

nucleoprotein: It is any protein which is structurally associated with nucleic acid (either DNA or RNA). The prototypical example is any of the histone class of proteins, which are identifiable on strands of chromatin. Telomerase, a RNP (RNA/protein complex), and protamines are also nucleoproteins.

nucleoside analogues: These are a range of antiviral products used to prevent viral replication in infected cells. The most commonly used is Acyclovir, although its inclusion in this category is uncertain, as it contains only a partial nucleoside structure, as the sugar ring is replaced by an open-chain structure.

nucleosides: They are glycosylamines consisting of a nucleobase bound to a ribose or deoxyribose sugar via a beta-glycosidic linkage. Examples of these include cytidine, uridine, adenosine, guanosine, thymidine and inosine. (See Appendix 4).

nucleosomes: They form the fundamental repeating units of eukaryotic chromatin, which are folded through a series of successively higher order structures to eventually form a chromosome.

nucleotides: These are the molecules that, when joined together, make up the structural units of RNA and DNA.

nucleus: It is a membrane bound organelle found in eukaryotic cells. It contains most of the cell's genetic material, organized as multiple long linear DNA molecules in complex with a large variety of proteins, such as histones, to form chromosomes. Its function is to maintain the integrity of these genes and to control the activities of the cell by regulating gene expression.

null mutation: It is a mutation in a gene that leads to its not being transcribed into RNA and/or translated into a functional protein product.

null-hypothesis: It is the hypothesis which is a test for possible rejection under the assumption that it is true.

nullisomy: It refers to diploid cell or organism lacking both members of a chromosome pair (2n -2).

nurse culture: It is a technique used in plant tissue culture in which planting a cell from a suspension culture is done on a raft of filter paper above a callus tissue piece (nurse tissue). The filter paper serves to prevent tissue union, but allows the flow of essential substances from the nurse to the isolated cell. In such a culture, a piece of callus is first placed on nutrient agar.

nutrient agar: It is a microbiological growth medium commonly used for routine cultivation of non-fastidious bacteria.

nutrient cycle: It is a pathway by which a chemical element or molecule moves through both biotic (biosphere) and abiotic (lithosphere, atmosphere, and hydrosphere) compartments of the earth.

nutrient deficiency: It refers to the absence or insufficiency of some factor needed for normal growth and development.

nutrient film technique (NFT): It is a hydroponic technique whereby a very shallow stream of water containing all the dissolved nutrients required for plant growth is recirculated past the bare roots of plants in a watertight gully, also known as channels.

nutrient gradient: It is a diffusion gradient of nutrients and gases that develops in tissues where only a portion of the tissue is in contact with the medium. Gradients are less likely to form in liquid media than in callus cultures.

O

object-relational database: It refers to the object databases combine the elements of object orientation and object-oriented programming languages with database capabilities. They provide more than persistent storage of programming language objects. Object databases extend the functionality of object programming languages (e.g., C++, Smalltalk, or Java) to provide full-featured database programming capability. The result is a high level of congruence between the data model for the application and the data model of the database. Object-relational databases are used in Bioinformatics to map molecular biological objects (such as sequences, structures, maps and pathways) to their underlying representations (typically within the rows and columns of relational database tables.) This enables the user to deal with the biological objects in a more intuitive manner, as they would in the laboratory, without having to worry about the underlying data model of their representation.

octoploid: It refers to a cell or organism having eight haploid sets of chromosomes in a body cell.

oestrogen or estrogen or estrogens: They are a group of steroid compounds, named for their importance in the estrous cycle, and functioning as the primary female sex hormone.

oestrous cycle: It comprises the recurring physiologic changes that are induced by reproductive hormones in most mammalian placental females. Humans undergo a menstrual cycle instead.

offspring (progeny): New individual organisms that result from the process of sexual or asexual reproduction.

ogive: A graph of a cumulative frequency distribution.

Okazaki fragment: It is a relatively short fragment of DNA (with an RNA primer at the 5' terminus) created on the lagging strand during DNA replication. The lengths of Okazaki fragments are between 1000 to 2000 nucleotides long in *E. coli*, and 100 to 200 nucleotides in eukaryotes.

OLA (oligonucleotide ligation assay): It is a rapid, sensitive, and specific method for the detection of known single nucleotide polymorphisms (SNPs).

oligomer: The term is used for short, single-stranded nucleic acid fragments, such as DNA or RNA, or similar fragments of analogues of nucleic acids such as peptide nucleic acid or morpholinos. It consists of less than five monomer units.

oligonucleotide: It is a short nucleic acid polymer, typically with twenty or fewer bases.

oligonucleotide-directed mutagenesis: It is a molecular biology technique in which a mutation is created at a defined site in a DNA molecule, usually a circular molecule known as a plasmid. In general, site-directed mutagenesis requires that the wild type gene sequence be known.

oligosaccharide: It is a saccharide polymer containing a small number of component sugars.

oncogene: It is a gene which when mutated or expressed at high levels, helps turn a normal cell into a tumour cell. Many abnormal cells undergo a programmed form of death.

oncomouse or Harvard mouse: It is a type of laboratory mouse that has been genetically modified using modifications designed by Philip Leder and Timothy A. Stewart.

one-tailed test: A hypothesis test in which there is one rejection region, i.e., we are concerned only whether the observed value deviates from the hypothesised value in one direction only.

ontogeny: It is the history of structural change in a unity, which can be a cell, an organism, or a society of organisms, without the loss of the organization that allows that unity to exist. It describes the origin and development of an organism from the fertilized egg to its mature form.

ontogeny: It refers to the development of an organism from the egg to maturity.

oocyte: The egg mother cell; it undergoes two meiotic divisions to form the egg cell.

oogenesis: The formation and growth of egg or ovum in an animal ovary.

oogonium: A germ cell of the female animal that gives rise to oocytes by mitotic division.

open continuous culture: It is a continuous culture in which inflow of fresh medium is balanced by outflow of a corresponding volume of culture. Cells are constantly washed out with the outflowing liquid. In a steady state, the rate of cell wash-out equals the rate of formation of new cells in the system.

open reading frame (ORF): It is a portion of an organism's genome which contains a sequence of bases that could potentially encode a protein. The start-points and end-points of a given ORF are not equivalent to the ends of the messenger RNA (mRNA), but the ends of the ORF are usually contained within the mRNA's sequence. In a gene, ORFs are located between the start-code sequence (initiation codon) and the stop-code sequence (termination codon).

operational definition: It is a demonstration of a process such as a variable, term, or object in terms of the specific process or a set of validation tests used to determine its presence and quantity.

operationalization: It is the process of defining a fuzzy concept so as to make the concept measurable in form of variables consisting of specific observations. In a wider sense, it refers to the process of specifying the extension of a concept.

operator: It refers to the region of DNA that is upstream from a gene or genes and to which one or more regulatory proteins (repressor or activator) binds to control the expression of the gene(s).

operon: It is a functioning unit of genomic material containing a cluster of genes under the control of a single regulatory signal or promoter.

opines: They are low molecular weight compounds found in plant crown gall tumours produced by the parasitic bacterium *Agrobacterium*.

ordinate and abscissa: In the construction of graphs, two simple lines are first drawn which cut each other at right angles. These lines are called axes. The horizontal line is called abscissa or x-axis, and the vertical line is called ordinate or y-axis. The point at which they cut each other is called the point of origin.

organ culture: It is a development from tissue culture methods of research. The organ culture is able to accurately model functions of an organ in various states and conditions by the use of the actual in vitro organ itself. Parts of an organ or a whole organ can be cultured in vitro.

organellar genes: They are the genes located on organelles outside the nucleus.

organelle: It is a specialized subunit within a cell that has a specific function, and is enclosed separately within its own lipid bilayer.

organic complex: It is a chemically undefined compound added to nutrient media to stimulate the growth, e.g., coconut milk, yeast extract, casein hydrolysate.

organic co-solvent: It is a compound used to dissolve some neutral organic substances, such as in media preparation. Organic co-solvents include alcohols (usually ethanol), acetone and dimethylsulphoxide.

organic evolution: It refers to the sequence of events involved in the evolutionary development of a species or taxonomic group of organisms.

organic farming: It is the form of agriculture that relies on crop rotation, green manure, compost, biological pest control, and mechanical cultivation to maintain soil productivity and control pests, excluding or strictly limiting the use of synthetic fertilizers and synthetic pesticides, plant growth regulators, livestock feed additives, and genetically modified organisms.

organic foods: They are the foods made according to certain production standards, usually in an organic farm.

organogenesis: It is the process by which the ectoderm, endoderm, and mesoderm develop into the internal organs of the organism.

organoid: It is a structure that resembles an organ.

organoleptic: It refers to any sensory properties of a product, involving taste, colour, odour and feel. Organoleptic testing involves inspection through visual examination, feeling and smelling of products.

origin of replication: It is a particular sequence in a genome at which replication is initiated.

ortholog: They are genes in different species that are evolved from a common ancestral gene by speciation. Normally, orthologs retain the same function in the course of evolution. Identification of orthologs is critical for reliable prediction of gene function in newly sequenced genomes. (See also Paralogs.)

osmic acid (osmium tetroxide): It is a chemical compound with the formula OsO_4. The compound is noteworthy for its many uses, despite the rarity of osmium.

osmolarity: It is the measure of solute concentration, defined as the number of osmoles (Osm) of solute per liter (L) of solution (osmol/L or Osm/L).

osmosis: It is the diffusion of water through a semi-permeable membrane from an area of high water potential (low solute concentration) to an area of low water potential (high solute concentration).

osmotic potential: The potential of water molecules to move from a hypotonic solution (more water, less solutes) to a hypertonic solution (less water, more solutes) across a semi-permeable membrane.

osmoticum: It refers to any substance that acts to supplement osmotic pressure in a plant or a culture of plant cells.

outbreeding depression: It refers to cases when offspring from crosses between individuals from different populations have lower fitness than the progeny from crosses between individuals from the same population.

outbreeding: It is the breeding of stocks or individuals that are not closely related. It is often used to counter the detrimental effects of continuous inbreeding.

overdominance: It is a condition in genetics where the phenotype of the heterozygote lies outside the phenotypical range of both homozygote parents.

overlapping clones: It refers to the collection of cloned sequences made by generating randomly overlapping DNA fragments with infrequently cutting restriction enzymes.

overlapping reading frames: It refers to the start codons in different reading frames that generate different polypeptides from the same DNA sequence.

overshoot: It refers to the crisis point of exhausting whatever resources were available.

ovulation: It is the process in a female's menstrual cycle by which a mature ovarian follicle ruptures and discharges an ovum.

ovule: It is the part of the reproductive organs in seed plants that consists of the nucellus, the embryo sac and integuments.

oxidation: It refers to changing a substance chemically by adding oxygen to it.

oxygen sensor (lambda sensor): It is an electronic device that measures the proportion of oxygen (O_2) in the gas or liquid being analyzed.

ozone: It refers to a three-atom oxygen molecule that in its gaseous state screens ultraviolet radiation. The protective ozone layer hangs 10 to 50 kilometres above the earth's surface.

P

packaging cell line: It refers to the cell line that is designed to produce viral particles that do not contain nucleic acid. After transfection of these cells with a full-size viral genome, fully infective viral particles are assembled and released.

packed cell volume (PCV) or hematocrit (HCT) or erythrocyte volume fraction (EVF): It is the proportion of blood volume that is occupied by red blood cells. It is normally about 48% for men and 38% for women. It is considered an integral part of a person's complete blood count results, along with haemoglobin concentration, white blood cell count, and platelet count.

PAGE (polyacrylamide gel electrophoresis): It is a technique used for the separation of deoxyribonucleic acid (DNA), ribonucleic acid (RNA), or protein molecules using an electric field applied to a gel matrix.

paired difference test: A hypothesis test of the difference between the sample means of two independent samples.

pairing (synapsis also called syndesis): It is the pairing of two homologous chromosomes that occurs during meiosis.

pair-rule gene: It is a type of gene involved in the development of segmented embryos of insects.

palaeontology: It is the study of prehistoric life, including organisms' evolution and interactions with each other and their environments.

palindrome: It refers to a word, phrase, number or other sequence of units that can be read the same way in either direction.

palindromic sequence: It is a segment of duplex DNA whose sequence is identical on each DNA strand. The sequence is the same when one strand is read left-to-right and the other strand is read right-to-left.

pAMP: Ampicillin-resistant plasmid.

PAN (peroxyacetyl nitrate): It refers to a pollutant found in photochemical smog. It irritates the eyes.

panmictic population: It is one where all individuals are potential partners. This assumes that there are no mating restrictions, neither genetic or behavioural, upon the population, and therefore all recombination is possible.

panmixis: It is a situation in which an individual is just as likely to mate with another randomly chosen individual as any other in the population.

paper raft technique: See nurse culture.

PAR (photosynthetically active radiation): It designates the spectral range (wave band) of solar radiation from 400 to 700 nanometres that photosynthetic organisms are able to use in the process of photosynthesis.

par gene: It is a gene found in bacterial and plant cells involved in the partition of plasmids in cells.

paracentric inversion: It is an inversion in a chromosome rearrangement in which a segment of a chromosome is reversed end to end. An inversion occurs when a single chromosome undergoes breakage and rearranges within itself. Paracentric inversions do not include the centromere, and both breaks occur in one arm of the chromosome. Pericentric inversions include the centromere, and there is a break point in each arm.

paraffin wax: It is a translucent, white, solid hydrocarbon with a low melting point. Paraffin is used as an embedding medium to support tissue for sectioning for light microscopy observation.

parahormone: It is a substance, not a true hormone, which has a hormone-like action in controlling the functioning of some distant organ.

parallel evolution: It is the development of a similar trait in different and not closely related species (that is in species of a different clade), but descending from the same ancestor.

paralog: They are genes related by duplication within a genome. Orthologs retain the same function in the course of evolution, whereas paralogs evolve new functions, even if these are related to the original one.

parameter: A measure which describes the characteristics of a population.

parameter: It is a quantity that serves to relate functions and variables using a common variable (often t) when such a relationship would be difficult to explicate with an equation. In different contexts, the term may have special uses.

parameters: They are user-selectable values, typically experimentally determined, that govern the boundaries of an algorithm or program. For instance, selection of the appropriate input parameters governs the success of a search algorithm. Some of the most common search parameters in bioinformatics tools include the stringency of an alignment search tool, and the weights (penalties) provided for mismatches and gaps.

parasexual cycle: It is a sequence involving heterokaryon formation, diploidization, and often resulting in the formation of recombinant nuclei. Unlike the sexual cycle, the parasexual cycle can occur at any point or continuously throughout the life cycle.

parasite: It is an organism that lives in a close relationship with another organism, its host, and harm it.

parasitism: It is a close association of two or more dissimilar organisms, where the association is harmful to at least one.

parasporal crystal: It refers to the tightly packaged insect pro-toxin molecules that are produced by the strains of *Bacillus thuringiensis* during the formation of resting spores.

PARs (proteolytically activated receptors): They are members of a larger family of seven transmembrane cell surface receptors that mediate cell activation events through heterotrimeric G proteins.

parthenocarpy: It is natural or artificially induced production of fruit without the fertilization of ovules. The fruit is therefore seedless.

parthenogenesis: It is a form of asexual reproduction found in females, where growth and development of embryos occurs without fertilization by a male.

partial digest: It is a restriction digest that has not been allowed to go to completion and thus contains pieces of DNA with some restriction endonuclease sites that have not yet been cleaved.

particle radiation: It refers to the radiation of energy by means of fast-moving subatomic particles. Particle radiation is referred to as a particle beam if all the particles are moving in the same direction, similar to a light beam.

particle swarm optimization (PSO): It is a computational method that optimizes a problem by iteratively trying to improve a candidate solution with regard to a given measure of quality. Such methods are commonly known as metaheuristics as they make few or no assumptions about the problem being optimized and can search very large spaces of candidate solutions. However, metaheuristics such as PSO do not guarantee an optimal solution is ever found.

parts per million (ppm): It is commonly used as a measure of small levels of pollutants in air, water, body fluids, etc. Parts per million is the mass ratio between the pollutant component and the solution, and it is defined as the units of any given substance per one million equivalent units, such as the weight units of solute per million weight units of solution (i.e., 1 ppm = 1 mg/L).

parturition: It refers to the childbirth, the process of delivering the baby and placenta from the uterus to the outside world.

Pascal triangle: A set of figures written in a triangular shape.

passage number: It refers to the number of times the cells in the culture have been subcultured.

passage time: Interval between successive subcultures.

passive immunity: It is the transfer of active humoral immunity in the form of readymade antibodies, from one individual to another.

passive immunization: It is used when there is a high risk of infection and insufficient time for the body to develop its own immune response, or to reduce the symptoms of ongoing or immunosuppressive diseases.

patent: It is a set of exclusive rights granted by the government to an inventor or their assignee for a limited period of time in exchange for a public disclosure of an invention.

paternal: Relating to or characteristic of a father or fatherhood; fatherly.

pathogen (infectious agent): It is a biological agent that causes diseases in the host.

pathogenesis: It is the development of a disease.

pathogen-free: It is a term used for laboratories or animals or plants that are guaranteed free of pathogens and enjoy the freedom from disease-causing organisms like bacteria, fungi, viruses, etc.

pathotoxin: It is a chemical of biological origin, other than an enzyme, that plays an important role in a plant disease.

pathovar: It is a bacterial strain or set of strains with the same or similar characteristics, which is differentiated at infrasubspecific level from other strains of the same species or subspecies on the basis of distinctive pathogenicity to one or more plant hosts.

pathways: Bioinformatics strives to define representations of key biological datatypes, algorithms and inference procedures, including sequences, structures, biological pathways and reactions. Representing and computing with biological pathways requires ontologies for representing pathway knowledge. User interfaces to these databases; physico-chemical properties of enzymes and their substrates in pathways; and pathway analysis of whole genomes including identifying common patterns across species and species differences.

pattern search (PS): It refers to a family of numerical optimization methods that do not require the gradient of the problem to be optimized, and PS can hence be used on functions that are not continuous or differentiable. Such optimization methods are also known as direct-search, derivative-free, or black-box methods.

pattern: It refers to the molecular biological patterns usually occurring at the level of characters making up the gene or protein sequence. A pattern language must be defined in order to apply different criteria to different positions of a sequence. In order to have position-specific comparison done by a computer, a pattern-matching algorithm must allow alternative residues at a given position, repetitions of a residue, exclusion of alternative residues, weighing, and ideally, combinatorial representation.

pBR322: It is a plasmid, and for a time was one of the most commonly used cloning vectors. It was the first artificial plasmid created in 1977, and was named eponymously after Bolivar and Rodriguez.

PCR: It is a technique to amplify a single or few copies of a piece of DNA across several orders of magnitude, generating thousands to millions of copies of a particular DNA sequence. The method relies on thermal cycling, consisting of cycles of repeated heating and cooling of the reaction for DNA melting and enzymatic replication of the DNA.

PDB: Protein Data Bank. It is resource for studying proteins.

pectin: It is a structural heteropolysaccharide contained in the primary cell walls of terrestrial plants. It was first isolated and described in 1825 by Henri Braconnot. It is produced commercially as a white to light brown powder, mainly extracted from citrus fruits, and is used in food as a gelling agent, particularly in jams and jellies. It is also used in fillings, sweets, as a stabilizer in fruit juices and milk drinks and as a source of dietary fibre.

pectinase: It is a general term for enzymes, such as pectolyase, pectozyme and olygalacturonase, commonly referred to in brewing as pectic enzymes.

pedigree: It is a table, chart or diagram recording the ancestry of an individual.

PEG (polyethylene glycol): It refers to an oligomer or polymer of ethylene oxide. They are liquids or low-melting solids, depending on their molecular weights.

penetrance: It refers to the proportion of individuals carrying a particular variation of a gene that also express an associated trait.

peptide bond: It is a chemical bond formed between two molecules when the carboxyl group of one molecule reacts with the amine group of the other molecule, thereby releasing a molecule of water (H_2O).

peptide vaccine: It is a short chain of amino acids that can induce antibodies against a specific infectious agent.

peptides: They are short polymers formed from the linking, in a defined order, of α-amino acids.

peptidyl transferase: It is an aminoacyltransferase as well as the primary enzymatic function of the ribosome which forms peptide links between the adjacent amino acids using tRNAs during the translation process of protein biosynthesis.

peptidyl-tRNA binding site (P-site): It is the site on a ribosome that hosts the tRNA to which an amino acid for the growing polypeptide chain is attached.

percentage bar diagram: In percentage bar diagrams, the length of the bars is kept equal to 100 and the divisions of the bar correspond to the percentage of different components.

pericentric inversion: It is an inversion that involves the centromere.

periclinal chimera: It refers to a plant composed of cells of two distinct species separated into distinctive zones.

periplasm: It is a space between the inner cytoplasmic membrane and external outer membrane of Gram-negative bacteria or the equivalent space outside the inner membrane of Gram-positive bacteria.

permanent wilting point (PWP): It is defined as the minimal point of soil moisture the plant requires not to wilt.

permeable: Substances through which fluids can penetrate or pass through.

permutation and combination: Permutation refers to different arrangement and combination refers to group.

persistence: Ability of an organism to remain in a particular setting for a period of time after it is introduced.

pesticide: It is a toxic chemical substance or mixture of substance intended for preventing, destroying, repelling or mitigating any pest.

petal: One of the parts of the flower that make up the corolla.

petite mutant: It is a class of yeast mutants, most studied in *Saccharomyces cerevisiae*.

Petri dish: It is a shallow glass or plastic cylindrical lidded dish that biologists use to culture cells. It is named after its inventor, Julius Richard Petri.

PFGE (pulsed-field electrophoresis): It is a gel electrophoresis technique used for the separation of DNA molecules. PFGE expands the range of resolution for DNA fragments by as much as 2 orders of magnitude.

pH: It is measure of the acidity or alkalinity of a solution. It is the negative logarithm (base 10) of molar concentration of dissolved hydrogen ions (H^+).

phage display: It refers to the technique in which phage are engineered to fuse a foreign peptide or protein with their capsid (surface) proteins and hence display it on their cell surfaces. The immobilized phage may then be used as a screen to see what ligands bind to the expressed fusion protein exhibited (displayed) on the phage surface.

phage therapy: It is the therapeutic use of bacteriophages to treat pathogenic bacterial infections.

Phage: It is any one of a number of viruses that infects bacteria. Bacteriophages are among the most common biological entities on the earth. The term is commonly used in its short form, *phage*.

phagemid (phasmid): It is a type of cloning vector developed as a hybrid of filamentous phage M13 and plasmids to produce a vector that can grow as a plasmid, and also be packaged as single-stranded DNA in viral particles. Phagemids contain an origin of replication (ori) for double-stranded replication, as well as an f1 ori to enable single-stranded replication and packaging into phage particles.

phagocytes: They are the white blood cells that protect the body by ingesting (phagocytosing) harmful foreign particles, bacteria, and dead or dying cells.

phagocytosis: It is the process of engulfing solid particles by the cell membrane to form an internal phagosome by phagocytes and protists. Phagocytosis is a specific form of endocytosis involving the vesicular internalization of solid particles, such as bacteria, and is therefore distinct from other forms of endocytosis such as the vesicular internalization of various liquids.

pharmacogenomics: It refers to the use of (DNA-based) genotyping in order to target pharmaceutical agents to specific patient populations. Genetic differences are known to affect responses to many types of drug therapy, and pharmacogenomics analysis serves to customize the use of pharmaceuticals for specific subgroups of patients. The rationale for this approach is that observed gene expression differences may correlate with, and explain, the differences in side effects and efficacy to drugs in humans.

pharmacophore: It refers to the three-dimensional spatial arrangement of atoms, substituents, functional groups, or chemical features that together are sufficient to describe the pharmacologically active components of a drug molecule or molecule series.

phenocopy: It is an organism or individual whose phenotype is identical to the another individual whose phenotype is determined by the genotype. In other words, the phenocopy environmental condition mimics the phenotype produced by a gene.

phenols: These are a class of chemical compounds consisting of a hydroxyl group (–OH) bonded directly to an aromatic hydrocarbon group. The simplest of the class is phenol (C_6H_5OH).

phenotype: It refers to any observable characteristic or trait of an organism.

phenylalanine: It is an α-amino acid with the formula $HO_2CCH(NH_2)CH_2C_6H_5$. The codons for L-phenylalanine are UUU and UUC.

pheromone: It is a secreted or excreted chemical factor that triggers a social response in members of the same species. Pheromones are chemicals capable of acting outside the body of the secreting individual to impact the behaviour of the receiving individual. There are *alarm pheromones*, *food trail pheromones*, *sex pheromones*, and many others that affect the behaviour or physiology. Their use among the insects has been particularly well documented. In addition, some vertebrates and plants communicate by using pheromones.

pheromone: These are the hormones capable of acting outside the body of the secreting individual to impact the behaviour of the receiving individual.

phloem: It is the living tissue in plants that carries organic nutrients.

phosphatase: It is an enzyme that removes a phosphate group from its substrate by hydrolysing phosphoric acid monoesters into a phosphate ion and a molecule with a free hydroxyl group.

phosphodiester bond: It is a group of strong covalent bonds between a phosphate group and two 5-carbon ring carbohydrates (pentoses) over two ester bonds.

phospholipase: It is an enzyme that hydrolyzes phospholipids into fatty acids and other lipophilic substances.

phospholipid: These are a class of lipids and are a major component of all cell membranes as they can form lipid bilayers.

phosphorolysis: It is the cleavage of a compound in which inorganic phosphate is the attacking group.

phosphorylation: It is the addition of a phosphate (PO_4) group to a protein or other organic molecule. Phosphorylation activates or deactivates many protein enzymes, causing or preventing the mechanisms of diseases such as cancer and diabetes.

photoautotrophs (autotrophs): They are the organisms, such as green plants or photoautotrophic bacteria, that synthesize organic materials using energy derived from sunlight in the process of photosynthesis.

photo-bioreactor: It is a bioreactor which incorporates some type of light source to provide photonic energy input into the reactor.

photochemical smog: It refers to air pollution produced by the reaction of hydrocarbons, nitrogen oxides, and sunlight. Automobile and industrial exhausts are major contributors.

photon: It is an elementary particle, the quantum of light and all forms of electromagnetic radiation.

photoperiod: The duration of an organism's daily exposure to light, considered especially with regard to the effect of such exposure on growth and development.

photoperiodicity: It is the physiological reaction of organisms to the length of day or night. It occurs in plants and animals.

photoperiodism: The response of an organism to changes in day length.

photophosphorylation: It is the production of ATP using the energy of sunlight. Only two sources of energy are available to living organisms: sunlight and oxidation-reduction (redox) reactions. All organisms produce ATP, which is the universal energy currency of life. In photophosphorylation, light energy is used to create a high-energy electron donor and a lower-energy electron acceptor. Electrons then move spontaneously from donor to acceptor through an electron transport chain.

photoreactivation: Reversal of the biological effects of ultraviolet radiation on cells by subsequent exposure to visible light.

photosynthate: A chemical product of photosynthesis.

photosynthesis: A chemical process that converts carbon dioxide into organic compounds, especially sugars, using the energy from sunlight, and occurs in plants, algae, and many species of bacteria, but not in archaea.

photosynthetic efficiency: It is the fraction of light energy converted into chemical energy during photosynthesis in plants and algae. Photosynthesis can be described by the simplified chemical reaction.

photosynthetically active radiation (PAR): It designates the spectral range (wave band) of solar radiation from 400 to 700 nanometres that photosynthetic organisms are able to use in the process of photosynthesis. This spectral region corresponds more or less with the range of light visible to the human eye. Photons at shorter wavelengths tend to be so energetic that they can be damaging to cells and tissues, but are mostly filtered out by the ozone layer in the stratosphere. Photons at longer wavelengths do not carry enough energy to allow photosynthesis to take place.

phototropism: It is a directional growth in which the direction of growth is determined by the direction of the light source. In other words, it is the growth and response to a light stimulus. Phototropism is most often observed in plants, but can also occur in other organisms such as fungi. The cells on the plant that are farthest from the light have a chemical called auxin that reacts when phototropism occurs.

phylogeny: It is the evolutionary development and history of a species or higher taxonomic grouping of organisms.

physical map of genes: It is a map showing physical locations on a DNA sequence, such as restriction sites and sequence-tagged sites.

phytochrome: It is a photoreceptor, a pigment that plants use to detect light. It is sensitive to light in the red and far-red region of the visible spectrum. Many flowering plants use it to regulate the time of flowering based on the length of day and night (photoperiodism) and to set circadian rhythms. It also regulates other responses including the germination of seeds, elongation of seedlings, the size, shape and number of leaves, the synthesis of chlorophyll, and the straightening of the epicotyl or hypocotyl hook of dicot seedlings.

phytoparasite: Any plant-parasitic organism.

phytopathogen: An organism that causes diseases in plants.

phytostat: The name adopted by Tulecke in 1965 for an apparatus designed for the semi-continuous chemostat culture of plant cells.

pie diagram: To represent qualitative data, different components or frequencies are said to be shown by means of sectors of a circle; the angles of the sectors are proportional to the respective measurements of the different components.

pigments: It is a material that changes the colour of reflected or transmitted light as a result of wavelength-selective absorption.

pinocytosis: It is commonly known as cell-drinking. It is a form of endocytosis in which small particles are brought into the cell suspended within small vesicles, which subsequently fuse with lysosomes to hydrolyze, or to break down, the particles.

pioneer: It refers to a plant species that establishes itself in a bare area like a recently levelled field until successional species move in.

pipette (Pasteur pipette): It is also called a pipet, pipettor or chemical dropper. It is a laboratory instrument used to transport a measured volume of liquid.

plant cell culture: It is the process by which cells are grown under controlled conditions. In practice, the term "cell culture" has come to refer to the culturing of cells derived from multicellular eukaryotes, especially animal cells. The historical development and methods of cell culture are closely interrelated to those of tissue culture and organ culture.

plant cell immobilization: It refers to the entrapment of plant cells in gel matrices in such a way that the cells are suspended in small drops of the material, which is then set or allowed to harden to make little carriers. Materials such as alginates, agar or polyacrylamide can be used. It is normally done in the preparation of artificial seeds.

plaque: A clear spot on an otherwise opaque culture plate of bacteria or cultured bacteria cells.

plasma cells: They are the short-lived antibody-producing cell derived from a type of leukocyte (white blood cell) called B cell. B cells differentiate into plasma cells that produce antibody molecules closely modelled after the receptors of the precursor B cell. Once released into the blood and lymph, these antibody molecules bind to the target antigen (foreign substance) and initiate its neutralization or destruction. Antibody production continues for several days or months, until the antigen has been overcome.

plasma membrane (cell membrane or plasmalemma): It is the cell's outer membrane made up of two layers of phospholipids with embedded proteins. It separates the contents of the cell from its outside environment, and regulates what enters and exits the cell.

Plasmid: It is an extrachromosomal, autonomous circular DNA molecule found in certain bacteria, capable of autonomous replication. Plasmids can transfer the genes between the bacteria and are important tools of transformation in genetic engineering. They exist in an autonomous state and are transferred independently of chromosomes.

plasmodesmata: They are microscopic channels which traverse the cell walls of plant cells and some algal cells, enabling transport and communication between them. Unlike animal cells, every plant cell is surrounded by a polysaccharide cell wall. Neighbouring plant cells are therefore separated by a pair of cell walls and the intervening lamella, forming an extracellular domain known as the apoplast. Although cell walls are permeable to small soluble proteins and other solutes, plasmodesmata enable direct, regulated, symplastic intercellular transport of substances between the cells. There are two forms of plasmodesmata: primary ones are formed during cell division, and secondary ones can form between mature cells.

plasmolysis: It is the process in plant cells where the plasma membrane pulls away from the cell wall due to the loss of water through osmosis. The reverse process, cytolysis, can occur if the cell is in a hypotonic solution resulting in a higher external osmotic pressure and net flow of water into the cell. Through observation of plasmolysis and deplasmolysis, it is possible to determine the tonicity of the cell's environment as well as the rate solute molecules cross the cellular membrane.

plastids: They are major organelles found in the cells of plants and algae. Plastids are the site of manufacture and storage of important chemical compounds used by the cell. Plastids often contain pigments used in photosynthesis, and the types of pigments present can change or determine the cell's colour.

plastoquinone: It is a quinone molecule involved in the electron transport chain in the light-dependent reactions of photosynthesis. Plastoquinone is reduced (accepts two protons(H^+) from the stromal matrix of the chloroplast, coupled to two electrons (e^-) from photosystem II, forming plastoquinol. It transports the protons to the lumen of thylakoid discs, while the electrons continue through the electron transport chain into the cytochrome b_6f protein complex.

plating efficiency: It is a measure of the number of colonies originating from single cells. It is a very sensitive test and is often used for determining the nutritional requirements of cells, testing serum lots, measuring the effects of growth factors, and toxicity testing.

platykurtic: A slightly peaked distribution.

pleiotropy: It occurs when a single gene influences multiple phenotypic traits. Consequently, a new mutation in the gene may have an effect on some or all traits simultaneously. This can become a problem when selection on one trait favours one specific version of the gene (allele), while the selection on the other trait favours another allele.

pleitropy: It refers to the multiple effects on an organism's phenotype due to a single gene or allele, e.g., the cytokines which can bind to multiple cellular receptors and effect growth and multiple immune pathways.

ploidy: It is the number of complete sets of chromosomes in a biological cell. In humans, the somatic cells that compose the body are diploid (containing two complete sets of chromosomes, one set derived from each parent), but sex cells (sperm and egg) are haploid. In contrast, tetraploidy (four sets of chromosomes) is a type of polyploidy and is common in plants, and not uncommon in amphibians, reptiles, and various species of insects.

point mutation (single base substitution): It is a type of mutation that causes the replacement of a single base nucleotide with another nucleotide of the genetic material, DNA or RNA. Often the term point mutation also includes insertions or deletions of a single base pair.

Poisson distribution: It is a discrete probability distribution. Poisson distribution is a limiting form of binomial distribution as 'n' moves towards infinity and 'p' moves towards zero and 'np' or mean remains constant.

Polar AUXIN transport: It is the regulated transport of plant hormone, auxin, in plants. It is suggested that it involves the components of the cytoskeleton, plasma membrane, and cell wall.

polar body: It is a cell structure found inside the ovum. Both animal and plant ova possess it. It is also known as a polar cell. The first polar body comprises one of the two products of meiosis-I, and it may not go through meiosis-II. The second polar body comprises one of the two products of meiosis-II.

polar mutation: A mutation that affects the transcription or translation of part of the gene or operon downstream of the mutant site. For example, nonsense mutations, frame shift mutations, and insertion sequence(IS)-induced mutations.

polar nuclei: In angiosperms, the two nuclei of the central cell of the female gametophyte; that fuse with a sperm nucleus to form the triploid (3n) endosperm nucleus.

polarity: The observed differentiation of an organism, tissue or cell into parts having opposed or contrasted properties or form. It is because of this property that the shoot and root development in plants occurs in the opposite direction.

pole cells: New cells that form at the posterior pole of the Drosophila egg that gives rise to the germ cells of the adult.

pollen culture: It is in vitro culture and germination of pollen grains.

pollen grain: It is a fine to coarse powder containing the **microgametophytes** of seed plants, which produce the male gametes (sperm cells).

pollination: It is the process by which pollen is transferred in plants, thereby enabling fertilization and sexual reproduction.

poly (A) polymerase: It is an enzyme that catalyzes the addition of adenine residues to the 3' end of pre-mRNAs to form the poly(A) tail.

poly-(A) tail: It is a set of sequences of adenine nucleotides that get added to the 3' end of some primary transcript messenger RNA molecules in eukaryotes during post-transcriptional processing. The added tail is believed to confer stability to the molecule.

polyacrylamide gel electrophoresis (PAGE): It is a technique used for the separation of deoxyribonucleic acid (DNA), ribonucleic acid (RNA), or protein molecules using an electric field applied to a gel matrix.

polyacrylamide gels: These are gels used as a separation matrix in electrophoresis of biomolecules, such as proteins or DNA fragments.

polyadenylation: It is the addition of a poly(A) tail to an RNA molecule.

polyclonal antibody (antisera): These are the antibodies obtained from different B cell resources. They are a combination of immunoglobulin molecules secreted against a specific antigen, each identifying a different epitope.

polyculture: It refers to the planting and growing of more than one species together, as natural world does it.

Polycyclic Aromatic Hydrocarbons (PAHs): It refers to over a hundred organic carcinogens produced by the incomplete burning of oil, coal, gas, tobacco, garbage, and even meat. Some are used in pesticides. Aside from causing cancer, their impact on humans is unknown, although test animals exposed to them lost the ability to reproduce.

polyembryony: It is a condition in which two or more embryos develop from a single fertilized egg, forming, what in humans is known as, identical twins.

polyethylene glycol (PEG) or carbowax: It is a polyether compound with many applications from industrial manufacturing to medicine.

polygene (multiple factor, multiple gene inheritance, or **quantitative gene):** It is a group of nonallelic genes that together influence a phenotypic trait. The loci or identities of the nonallelic genes are frequently unknown to biologists, even though they are known to exist. The genes that contribute to type 2 diabetes are thought to be mostly polygenes.

polygenic inheritance: It occurs when one characteristic is controlled by two or more genes (usually by many different genes) at different loci on different chromosomes. These genes are described as polygenes.

polylinker or multiple cloning site (MCS): A short DNA sequence containing several restriction enzyme recognition sites.

polymer: It is a large molecule (macromolecule) composed of repeating structural units typically connected by covalent chemical bonds. While *polymer* in popular usage suggests plastic, the term actually refers to a large class of natural and synthetic materials with a wide variety of properties, including properties typically associated with plastics.

polymerase: It is an enzyme whose central function is associated with polymers of nucleic acids such as RNA and DNA. The primary function of a polymerase is the polymerization of new DNA or RNA against an existing DNA or RNA template in the processes of replication and transcription.

polymerization: It is a process of reacting monomer molecules together in a chemical reaction to form three-dimensional networks or polymer chains. There are many forms of polymerization and different systems to categorize them.

polymery: It is the phenomenon whereby a number of genes at different loci can act together to produce a single effect.

polymorphism (gene): The occurrence of two or more alleles at a locus in a population.

polymorphism: It refers to the expression of two or more of a gene's alleles (varieties) in a population. Human hair and skin colour are polymorphous.

polynucleotide: It is an organic polymer molecule composed of 13 or more nucleotide monomers covalently bonded in a chain. DNA (deoxyribonucleic acid) and RNA (ribonucleic acid) are examples of polynucleotides with distinct biological function.

polypeptide: A linear molecule composed of two or more amino acids linked by peptide bonds. They are called dipeptides, tripeptides and so forth, according to the number of amino acids present.

polyploids: Cells and organisms having more than two paired (homologous) sets of chromosomes.

polysaccharide capsules: It is a carbohydrate covering with antigenic specificity that are present on some types of bacteria.

polysaccharide: These are polymeric carbohydrate structures formed of repeating units (either mono- or disaccharides) joined together by glycosidic bonds. These structures are often linear, but may contain various degrees of branching. Polysaccharides are often quite heterogeneous, containing slight modifications of the repeating unit.

polytene chromosomes: These are giant chromosomes that form when multiple rounds of replication produce many sister chromatids that remain synapsed together. In addition to increasing the volume of the cell's nuclei and causing cell expansion, polytene cells may also have a metabolic advantage as multiple copies of genes permits a high level of gene expression.

polyunsaturates: They are a type of fat or oil containing a high proportion of triglyceride molecules whose fatty acid chains contain several double bonds. By contrast, the fatty-acid chains of triglycerides in saturated fats (such as lard) contain only single bonds. Medical evidence suggests that polyunsaturated fats, used widely in margarines and cooking fats, are less likely to contribute to cardiovascular disease than saturated fats, but there is also some evidence that they may have adverse effects on health.

polyvalent vaccine: It is a vaccine prepared from cultures or antigens of more than one strain or species.

polyvinylpyrrolidone (PVP): It is a water-soluble polymer made from the monomer *N*-vinylpyrrolidone. Occasionally, it is used as a constituent of plant tissue culture isolation media. PVP is of variable molecular weight and has antioxidant properties, so it is used to prevent oxidative browning of explanted tissues. PVP is less frequently used as an osmoticum in culture media.

population density: It is a measurement of population per unit area or unit volume. It is frequently applied to living organisms, and particularly to humans. It is a key geographic term.

population genetics: It is the branch of genetics that deals with the frequencies of alleles and genotypes in breeding populations.

Population: It is a local group of organisms belonging to the same species and interbreeding.

population: The total number of individual observations from which inferences are to be made at a particular time.

porcine endogenous retrovirus (PERV): It is the provirus of aporcine retrovirus.

position effect: It is the effect on the expression of a gene when its location in a chromosome is changed, often by translocation.

positional candidate gene: It is a gene known to be located in the same region as a DNA marker that has been shown to be linked to a single-locus trait or a quantitative trait locus, and whose function suggests that it could be a source of genetic variation in the trait in question.

positional cloning: It is a technique used in genetic screening to identify specific areas of interest in the genome, and then determine what they do. This type of genetic screening is sometimes referred to as reverse genetics, because researchers start by figuring out where a gene is, and then they determine what it does, in contrast with methods which start by determining the function of a gene and then finding it in the genome.

positive control system: It is a mechanism in which the regulatory protein(s) is required to turn on gene expression.

positive correlation: The values of the two variables move in the same direction.

post-replication repair: It is the process of repairing DNA that occurs soon after the replication, as DNA damage prevents the normal enzymatic synthesis of DNA by the replication fork. At damaged sites in the genome, both prokaryotic and eukaryotic cells utilize a number of post-replication repair (PRR) mechanisms to complete DNA replication.

post-translational modification (PTM): It is the chemical modification of a protein after its translation. It is one of the later steps in protein biosynthesis for many proteins.

ppm: parts per million.

precocious germination: It refers to seed germination before the fruit has released seed. The seeds of the green apple commonly germinate in this manner.

precursor mRNA(pre-mRNA): It is an immature single strand of messenger ribonucleic acid (mRNA). pre-mRNA is synthesized from a DNA template in the cell nucleus by transcription. pre-mRNA comprises the bulk of **heterogeneous nuclear RNA (hnRNA)**. The term hnRNA is often used as a synonym for pre-mRNA, although strictly speaking, hnRNA may include nuclear RNA transcripts that do not end up as cytoplasmic mRNA.

pregnancy test: It is a method to determine whether or not a woman is pregnant.

pressure potential: It is the hydrostatic pressure to which water in a liquid phase is subjected. It was known formerly as wall pressure or turgor pressure. In a turgid plant cell, pressure potential is usually positive, but the pressure potential of xylem in a transpiring plant, which is under considerable tension, will be negative.

pre-transplant: It is one of the steps in tissue culture or micropropagation. It is the third and the stage of rooting and hardening stage prior to transfer to soil.

preventive immunization (vaccination): It is the administration of antigenic material to produce immunity against a disease. Vaccines can prevent or ameliorate the effects of infection by many pathogens.

Pribnow box (Pribnow-Schaller box): It is the sequence *TATAAT* of six nucleotides (thymine-adenine-thymine, etc.) that is, an essential part of a promoter site on DNA for transcription to occur in bacteria.

primary antibody: They are antibodies raised against an antigenic target of interest and are typically unconjugated. They recognize and bind with high affinity and specificity to unique epitopes across a broad spectrum of biomolecules and they are available as high specificity monoclonal antibodies and/or as polyclonal antibodies.

primary cell: It is a cell or cell line taken directly from a living organism, which is not immortalized.

primary culture: It is a cell or tissue culture started from material taken directly from an organism, as opposed to that taken from an explant of an organism.

primary germ layers: These are germ layers produced during the initial development of organism from the zygote.

primary immune response: It is the immune response to the first challenge by a particular antigen. Usually less extensive than the secondary immune response, being slower and shorter-lived with smaller amounts of lower affinity antibody being produced.

primary transcript: It is an RNA molecule that has not yet undergone any modification after its synthesis. For example, a precursor messenger RNA (pre-mRNA) is a primary transcript that becomes a messenger RNA (mRNA) after processing, and a primary microRNA transcript (pri-miRNA) becomes a microRNA (miRNA) after processing.

primer DNA polymerase: It is a DNA polymerase that provides primers for the DNA polymerization. Unlike RNA polymerase, DNA polymerase is unable to initiate the de novo synthesis of a polynucleotide chain.

primer walking: It is a sequencing method of choice for sequencing DNA fragments between 1.3 and 7 kilobases. Such fragments are too long to be sequenced in a single sequence read using the chain termination method. This method works by dividing the long sequence into several consecutive short ones. The DNA of interest may be a plasmid insert, a PCR product or a fragment representing a gap when sequencing a genome. The term "primer walking" is used where the main aim is to sequence the genome. The term "chromosome walking" is used instead when we know the sequence but do not have a clone of a gene.

primer: It is a strand of nucleic acid that serves as a starting point for DNA synthesis. It is required because the enzymes that catalyze replication, DNA polymerases, can only add new nucleotides to an existing strand of DNA. The polymerase starts replication at the 3'-end of the primer, and copies the opposite strand.

primordium: It is a group of cells or an organ or tissue in its earliest recognizable stage of development.

primosome: It is a complex of seven proteins: DnaG primase, DnaB helicase, DnaC helicase assistant, DnaT, PriA, Pri B, and PriC. The primosome creates RNA primers on a single-stranded DNA during DNA replication.

prion (proteinaceous infectious particle): It is an infectious agent composed primarily of protein. Prions cause a number of diseases in a variety of mammals.

probability distribution: A graphical representation which shows the possible outcome of a series of experiments and their respective probability.

probability tree: A graphical representation which shows the possible outcome of a series of experiments and their respective probability.

probability: Probability is the likelihood of occurrence of an event.

proband (propositus): It is a term used often in medical genetics and other medical fields to denote a particular subject (person or animal) being studied or reported on. On pedigrees, the proband is noted with an arrow and the box (male), or circle (female) shaded accordingly. It is important to denote the proband, so that the relationship to other individuals can be seen and patterns established.

probe (DNA): It a single-stranded DNA molecule used in laboratory experiments to detect the presence of a complementary sequence among a mixture of other singled-stranded DNA molecules.

profile: Profiles are usually derived from multiple alignments of sequences with a known relationship, and consist of tables of position-specific scores and gap-penalties. Each position in the profile contains scores for all of the possible amino acids, as well as one penalty score for opening and one for continuing a gap at the specified position. Attempts have been made to further improve the sensitivity of the profile by refining the procedures to construct a profile starting from a given multiple alignment. Other representations for sequence domains or motifs do not necessarily require the presence of a correct and complete multiple alignment, such as hidden Markov models.

progenesis: It refers to sexual maturity reached while still in the larva stage.

progeny testing: It is used in the breeding of both plants and animals, but is commercially important in animal breeding to determine the value of an animal in terms of the sex-limited characters of its offspring, for example, milk or egg production in females. A bull, for example cannot be assessed for milk production, however the performance of its female offspring can be used to determine the use of animal for future crosses, provided of course the character has a high heritability. A progeny test is performed by mating the male with a number of females with proven performance. The average performance of the offspring is then found, giving a measure of the male's respective value to the breeder.

progeny: It is the product of reproduction, a new organism produced by one or more parents.

progesterone: It is a steroid hormone involved in the female menstrual cycle, pregnancy and embryogenesis of humans and other species. Progesterone belongs to a class of hormones called progestogens, and is the major naturally occurring human progestogen.

prokaryotes: They are a group of organisms that lack a cell nucleus (= karyon), or any other membrane-bound organelles. They differ from the eukaryotes, which have a cell nucleus. Most are unicellular, but a few prokaryotes such as myxobacteria have multicellular stages in their life cycles.

prolactin: It is a peptide hormone associated with lactation.

proliferation: Increase in the number of cells or organisms by frequent and repeated reproduction.

promoter sequence: A regulatory DNA sequence that initiates the expression of a gene.

promoter: A nucleotide sequence of DNA to which RNA polymerase binds and initiates transcription.

pro-nuclear micro-injection: The initial method of transgenesis in animals, involving injecting many copies of a particular gene into one of the two pro-nuclei of a fertilized ova. It is characterised by a very low success rate.

pro-nucleus: It is the nucleus of a sperm or an egg cell during the process of fertilization.

proofreading(DNA): The enzymatic scanning of newly-synthesized DNA for structural defects, such as mismatched base pairs. It is a process effected by DNA polymerase.

propagation (plant): It is the process of artificially or naturally distributing plants.

propagule: It is any plant material used for the purpose of plant propagation.

prophage: It is a phage genome inserted as part of the linear structure of DNA chromosome of a bacterium.

prophase: It is a stage of mitosis in which the chromatin condenses into a highly ordered structure called chromosome in which the chromatin becomes visible.

protamines: They are small, arginine-rich, nuclear proteins that replace histones late in the haploid phase of spermatogenesis, and are believed to be essential for sperm head condensation and DNA stabilization.

protease: It is an enzyme that hydrolyzes proteins, cleaving the peptide bonds that link amino acids in protein molecules.

protein crystallization: Process of transforming proteins into crystals when placed in the appropriate conditions for determining three-dimensional structure of a protein. In order to crystallize a protein, the purified protein undergoes slow precipitation from an aqueous solution.

protein engineering: It is the process of developing useful or valuable proteins. A detailed knowledge of protein structure and function, as well as advancements in high-throughput technology, may greatly expand the capabilities of protein engineering.

protein ID: It is an accession number given to the author when a scientist submits a studied protein structure to PDB (Protein Data Bank). For example, 2H5T is the ID given by PDB to Dr. Jayarama Reddy and his associates for the protein Magee-1.

protein kinase: It is a kinase enzyme that modifies other proteins by chemically adding phosphate groups to them.

protein sequencing: It is the process of determining the amino acid sequences of the protein or its constituent peptides. It also determines what conformation it adopts and whether it is complexed with any non-peptide molecule.

protein synthesis: It is a process in which the cells build proteins. The term is sometimes used to refer only to protein translation, but more often it refers to a multi-step process, beginning with amino acid synthesis and transcription of nuclear DNA into messenger RNA, which is then used as input to translation.

protein: These are organic compounds made of amino acids arranged in a linear chain and folded into a globular form.

proteolysis: It is enzymatic degradation of proteins by intramolecular digestion or cellular enzymes called proteases.

proteome: It refers to the entire protein complement of a given organism.

proteomics: It refers to the study of the proteome. Typically, the cataloging of all the expressed proteins in a particular cell or tissue type, obtained by identifying the proteins from cell extracts using a combination of 2D gel electrophoresis and mass spectrometry. The large-scale analysis of the protein composition and function.

protoclone: It is a regenerated plant derived from protoplast culture or single colony derived from protoplasts in culture.

protocol: It is a set of guidelines or rules.

protocorm like bodies (plbs): These are the small swollen tubers, mostly undifferentiated cells. May evolve from orchid cells in tissue-culture. Can be made to produce shoots in vitro with cytokinins that finally become small clone plants.

protocorm: It is a mass of cells produced when a seed germinates; the first stage of germination. This term is usually used in discussing orchid seed germination.

protogyny: It refers to a condition in hermaphroditic or dioecious organisms in which the female reproductive structures mature before the male structures.

proto-oncogene: It is a normal gene that can become an oncogene due to mutations or increased expression.

protoplasm: It is the living content of a cell surrounded by plasma membrane. This term is not commonly used in modern cell biology. Protoplasm is composed of a mixture of small molecules such as ions, amino acids, monosaccharides and water, and macromolecules such as nucleic acids, proteins, lipids and polysaccharides. In eukaryotes, the protoplasm surrounding the cell nucleus is known as cytoplasm, and that inside the nucleus as the nucleoplasm. In prokaryotes, the material inside the plasma membrane is bacterial cytoplasm, while in Gram-negative bacteria the region outside the plasma membrane but inside the outer membrane is the periplasm.

protoplast culture: It is one of the important methods of plant tissue culture. It involves the isolation and culture of plant protoplasts by mechanical means or enzymatic digestion of plant tissues or organs, or cultures derived from these. Protoplasts are utilized for selection or hybridization at the cellular level and for a variety of other purposes.

protoplast fusion: It refers to a technique by which two protoplasts are fused to form hybrid cells that can grow into mature hybrid organisms. It is usually performed on plants.

protoplast: These are spherical and smaller than the elongate, angular-shaped and often vacuolated cells from which they have been released. In general, it is referred to the bacterial or plant cell in which the relatively rigid wall has been removed either chemically or enzymatically, leaving its cytoplasm enveloped by only a delicate peripheral membrane.

prototroph: It refers to a microorganism that has the ability to synthesize all of its amino acids, nucleic acids, vitamins, and other cellular constituents from inorganic nutrients.

pro-toxin: A chemical compound that is a precursor to a toxin.

protozoan: A microscopic, single-cell organism.

provirus: A virus genome that has integrated itself into the DNA of a host cell.

pseudogene: These are dyefunctional relatives of known genes that have lost their protein-coding ability or are otherwise no longer expressed in the cell.

pseudomonas: It is a common Gram-negative bacterial genus.

psychrophile (cryophiles): They are extremophilic organisms that are capable of growth and reproduction in cold temperatures, ranging from –15°C – +10°C. They are found in pockets of very salty water (brine) surrounded by sea ice. Psychrophiles use a wide variety of metabolic pathways, including photosynthesis, chemoautotrophy, and heterotrophy, and form robust, diverse communities. They are characterized by lipid cell membranes chemically resistant to the stiffening caused by extreme cold, and often create protein 'anti-freezes' to keep their internal space liquid and protect their DNA even in temperatures below water's freezing point. Hence, they are of great importance for the research in molecular biology. Species of examples *Arthrobacter* and *Psychrobacter* are some examples.

pUC: It is a widely used expression plasmid, containing a galactosidase gene. It is the abbreviated form of plasmid of University of California.

pulsed-field gel electrophoresis (PFGE): It is a technique which is relatively similar to performing a standard gel electrophoresis except that instead of constantly running the voltage in one direction, the voltage is periodically switched among three directions—one that runs through the central axis of the gel, and two that run at an angle of 120 degrees either side.

punctuated equilibrium: It is a theory in evolutionary biology which proposes that some sexually reproducing species will experience little evolutionary change for most of their geological history, remaining in an extended state called stasis.

pure culture (axenic culture): It is the growth and maintenance of a single species in isolation, free from foreign or contaminating species. Isolation is usually achieved by growing the species in an environment that was previously sterilized, and was thereby rid of contaminating organisms. Since, from a practical viewpoint, the contaminating organisms usually encountered are microorganisms, axenic cultures, whether of invertebrates or vertebrates, are often referred to as germ free.

pure line: It is a breed or strain of animals or plants that maintains a high degree of consistency in certain characters as a result of inbreeding for generations.

purine: It is a heterocyclic aromatic organic compound, consisting of a pyrimidine ring fused to an imidazole ring. Purines, including substituted purines and their tautomers, are the most widely distributed nitrogen-containing heterocycle in nature.

pyrimidine: It is a heterocyclic aromatic organic compound similar to benzene and pyridine, containing two nitrogen atoms at positions 1 and 3 of the six-member ring. It is isomeric with two other forms of diazine.

pyrogen: It is a substance inducing fever. Similarly, a mediator of fever, produced by polymorphonuclear leukocytes, monocytes and macrophages is called an endogenous pyrogen.

Python: It is a general-purpose high-level programming language whose design philosophy emphasizes code readability.

Q

quadrant: It is an instrument that is used to measure angles up to 90°.

quadrat: It is a square (of either metal, wood, or plastic) used in ecology to isolate a sample, usually about $1m^2$ or $0.25\ m^2$. The quadrat is suitable for sampling plants, slow-moving animals (such as millipedes and other insects), and some aquatic organisms.

quadriplegia: It is also known as tetraplegia and is a paralysis caused by illness or injury to a human that results in the partial or total loss of use of all of their limbs and torso.

Quadrivalent Human Papillomavirus Vaccine(QHPV): It is a vaccine against the most common sexually transmitted virus.

quadruplets: They are the four children born in one pregnancy.

quadrupole mass spectrometer: It consists of four precisely straight and parallel rods so arranged that the beam of ions from the ionization source of spectrometer is directed axially between them. A voltage comprising a DC component and a radio frequency component is applied between the adjacent rods, opposite rods being electrically connected.

qualitative characters: Individuals comprising the material under consideration are distinguished by some quality.

quantitative characters: Individuals are distinguished by measurement.

quantitative genetics: It is the area of genetics concerned with the study of continuous traits (such as height or weight) and their underlying mechanisms. It is effectively an extension of simple Mendelian inheritance in that the combined effect of many underlying genes results in a continuous distribution of phenotypic values.

quantitative inheritance: It is a type of inheritance in which each dominant allele of all the different genes contribute the same amount to the phenotype. This type of inheritance of measurable traits (height, weight, colour intensity, etc.) depends on the cumulative action of many genes.

quantitative trait locus QTLs): They are stretches of DNA closely linked to the genes that underlie the trait in question. QTLs can be molecularly identified (for example, with AFLP) to help map regions of the genome that contain genes involved in specifying a quantitative trait. This can be an early step in identifying and sequencing these genes.

quantitative trait: It is a phenotype that is influenced by multiple genes. It is a measurable trait that shows continuous variation.

quantum speciation: It is the rapid formation of new species, primarily by genetic drift.

quarantine: It is voluntary or compulsory isolation, typically to contain the spread of something considered dangerous, often but not always, disease. It was originally used for keeping a person or a living organism in isolation for a period after arrival to allow disease symptoms to appear, if there was any disease present. It is now used for regulations restricting the sale or shipment of living organisms to prevent disease or pest invasion of an area.

quark: It refers to the particle that makes up protons and neutrons.

query (sequence): It refers to the DNA, RNA of protein sequence used to search a sequence database in order to identify close or remote family members (homologs) of known function, or sequences with similar active sites or regions (analogs), from whom the function of the query may be deduced.

quiescence: It is the non-active state of a seed in which the only requirement for seed germination is water and oxygen.

R

race: It is any inbreeding group, including taxonomic subgroups such as subspecies, taxonomically subordinate to a species and superordinate to a sub-race and marked by a pre-determined profile of latent factors of hereditary traits.

raceme: It is a type of inflorescence in which the main axis is elongated but the flowers are borne on pedicels that are about equal in length.

radical: It is that portion of plant embryo which develops into primary or seed root.

radicle: It refers to the root of a plant embryo; the first part of a seedling to emerge during germination.

radioactive isotope: It is an unstable isotope that emits ionizing radiation.

raft culture: In tissue culture it is a method of growing tissues suspended from rafts or floats. The term is sometimes used to describe a method of hanging culture. Skin is grown by raft culture. In an organotypic raft culture, normal human keratinocytes stratify and form a differentiated epithelium reproducing different layers of skin.

ramet: It is an individual member of a clone.

random amplified polymorphic DNA (RAPD): It is a type of PCR reaction, but the segments of DNA that are amplified are random. The scientist performing RAPD creates several arbitrary, short primers (8-12 nucleotides), and then proceeds with the PCR using a large template of genomic DNA, hoping that the fragments will amplify.

random genetic drift: It refers to the changes in allelic frequency due to sampling error. Changes in allele frequency result because the genes appearing in offspring are not a perfectly representative sampling of the parental genes.

random mutagenesis: It is a non-directed change of one or more nucleotide pairs in a DNA molecule.

random primer method: It is a method of labelling DNA in vitro.

random sampling: Random sampling is one in which each and every item of the population has the same probability of being included in the sample. It completely depends on the element of chance.

random selection: It is one in which each unit of the population has the same chance of being included in the sample.

randomization: Randomization is a required condition for an experimental design which is necessary to obtain a valid estimate of the error variation. The process of randomization will ensure that the various soil conditions represented in the field will have an equal chance of being used in the experiment.

range: The difference between the largest and the smallest value of the distribution of data.

Rat (*Rattus norvegicus*): The rat is a principal model organism to link function to genes. The large number of inbred rat models and the vast amount of data available for the rat provide important strengths for the study of human health and disease. In some instances, specific aspects of human disease are duplicated well only in the rat, making these animals a unique resource for studying and identifying genetic pathways relevant to human disease. The rat is also a model of choice for many physiological studies related to cardiac and vascular function, pulmonary circulation, metabolism, neurological control, age- and gender-related differences, and studies related to hypertension and signal transduction.

ratbot (robo-rat): It is a rat with electrodes implanted in the medial forebrain bundle (MFB) and sensorimotor cortex of its brain. They were developed in 2002 by Sanjiv Talwar and John Chapin at the State University of New York Downstate Medical Center. The rats wear a small electronics backpack containing a radio receiver and electrical stimulator. The rat receives remote stimulation in the sensorimotor cortex via its backpack that causes the rat to feel a sensation in its left or right whiskers, and stimulation in the MFB that is interpreted as a reward or pleasure.

rational drug design (structure-based drug design): It refers to the development of drugs based on the 3-dimensional molecular structure of a particular target.

raw data: Information before it is arranged or analysed by statistical methods.

reading frame: It is a contiguous and non-overlapping set of three-nucleotide codons in DNA or RNA. There are 3 possible reading frames in an mRNA strand and six in a double-stranded DNA molecule due to the two strands from which transcription is possible. This leads to the possibility of overlapping genes, and there may be many of these in bacteria.

reading frame: It refers to the sequence of codons beginning with an initiation codon and ending with a termination codon, typically of at least 150 bases (50 amino acids) coding for a polypeptide or protein chain (see ORF and URF).

read-through: Transcription or translation beyond the normal termination signals in DNA or mRNA respectively.

rec A: It is an *E. coli* protein essential for the repair and maintenance of DNA. It has multiple activities, all related to DNA repair. In the bacterial SOS response, it has a co-protease function in the autocatalytic cleavage of the LexA repressor and the λ repressor. Since it is a DNA-dependent ATPase, RecA contains an additional site for binding and hydrolyzing ATP. RecA associates more tightly with DNA when it has ATP bound than when it has ADP bound.

recalcitrant seed: It is a seed that cannot survive desiccation, and germinates instantly. Its storage life is usually only a few days, rarely a few weeks. Most climaxtropical rainforest tree species produce recalcitrant seeds, which must be stored as growing plants rather than the seed.

receptor: It is a molecular structure or site on the surface or interior of a cell that binds with substances such as hormones, antigens, drugs, or neurotransmitters.

receptor-binding screening: It is one technique of biotechnology for discovering conventional drugs. The method relies on the fact that many drugs act by binding to specific proteins (receptors) on or in the cells. These proteins usually bind to hormones or other cells, and control the cell's behaviour, although they may be enzymes or structural elements of the cell. The drug interferes with the normal role of the protein.

recessive oncogene: A single copy of this gene is sufficient to suppress cell proliferation; loss of both copies of the gene contributes to cancer formation.

recessive: It is a term describing an allele whose effect with respect to a particular trait is not evident in heterozygotes.

reciprocal crosses: It is a breeding experiment designed to test the role of parental sex on a given inheritance pattern. All parent organisms must be true breeding to properly carry out such an experiment.

reciprocating shaker: It is a platform shaker used for agitating culture flasks, with a back and forth action at variable speeds. It is uses in laboratories for incubation and moderating rooms at ambient temperatures between +10 and +50°C, in medical diagnostics or analysis as well as for growing microorganisms, cell and tissue cultures.

recognition sequence (recognition site): It is any DNA-binding protein motif that exhibits binding specificity. It refers to the DNA sequence to which the domain is specific, or a location on a nucleic acid or protein to which a specific ligand binds.

recombinant DNA: It is a form of artificial DNA that is created by combining two or more sequences that would not normally occur together. In terms of genetic modification, it is created through the introduction of relevant DNA into an existing organismal DNA, such as the plasmids of bacteria, to code for or alter different traits for a specific purpose, such as antibiotic resistance. It differs from genetic recombination in that it does not occur through natural processes within the cell, but is engineered.

recombinant protein: A protein produced using recombinant DNA.

recombinant vaccine: A vaccine produced using recombinant DNA.

recombination (genetic): It is a process by which a molecule of nucleic acid (usually DNA; but can also be RNA) is broken and then joined to a different DNA molecule. Recombination can occur between similar molecules of DNA, as in homologous recombination, or dissimilar molecules of DNA as in non-homologous end joining. Recombination is a common method of DNA repair in both bacteria and eukaryotes.

recursion: It refers to the algorithmic procedure whereby an algorithm calls on itself to perform a calculation until the result exceeds a threshold, in which case the algorithm exits. Recursion is a powerful procedure with which to process data and is computationally quite efficient.

re-differentiation: It is a process by which a group of once differentiated cells return to their original specialized form.

redox (reduction-oxidation): It refers to a chemical transformation in which electrons are removed from one substance (oxidation) and added to another (reduction). The reductant transfers electrons to the oxidant. Redox reactions transfer biological energy. In photosynthesis, carbon dioxide is reduced sugars; in respiration, sugars are oxidized to make carbon dioxide and water.

reduction division: It is a nuclear division that produces daughter nuclei each having one-half as many centromeres as the parental nucleus. The first meiotic division reduces the number of chromosomes and centromeres to half that of the original cell.

regeneration: It is a process in which a lost or damaged part of regrows so that the original function is restored.

regression analysis: This technique is concerned with measuring the probable form of relationship between the two variables.

regression equations: The two equations based on two regression lines are called regression equations.

regression line: A line fitted to a set of data point to estimate the relationship between two variables.

regression: The statistical method which helps us to estimate the unknown value of one variable from the known value of the related variable.

regulator gene: It is a gene involved in controlling the expression of one or more other genes. A regulator gene may encode a protein, or it may work at the level of RNA, as in the case of genes encoding microRNAs.

regulatory gene: It refers to the DNA sequence that functions to control the expression of other genes by producing a protein that modulates the synthesis of their products (typically by binding to the gene promoter). (cf. Structural gene).

rejuvenation: It is the reversal of aging and thus requires a different strategy, namely repair of the damage that is associated with aging or replacement of damaged tissue with new tissue. In some plants, such as diatoms, rejuvenating spores are produced.

Relational Database Management Systems (RDBMS): It refers to the software system that includes a database architecture, query language, and data loading and updating tools and other ancillary software that together allow the creation of a relational database application.

relational database: It refers to the database that follows E.F. Codd's 11 rules, a series of mathematical and logical steps for the organization and systemization of data into a software system that allows easy retrieval, updating, and expansion. An RDBMS stores data in a database consisting of one or more tables of rows and columns. The rows correspond to a record (tuple); the columns correspond to attributes (fields) in the record. In an RDBMS, a view, defined as a subset of the database that is the result of the evaluation of a query, is a table. RDBMSs use Structured Query Language (SQL) for data definition, data management, and data access and retrieval. Relational and object-relational databases are used extensively in bioinformatics to store sequence and other biological data.

relative frequency distribution: The display of a data set that shows the fraction or percentage of the total data set that falls into each of a set of mutually exclusive classes.

relaxed circle: It is also called nicked plasmid. During, extraction of plasmid DNA from the bacterial cell, one strand of the DNA becomes nicked. This relaxes the torsional strain needed to maintain supercoiling, and producing the familiar form of plasmid.

release factors: It is a protein that allows the termination of translation by recognizing the termination codon or stop codon in a mRNA sequence.

re-naturation: The process by which proteins or complementary strands of nucleic acids reform their native conformations.

rennin: It is also called chymosin, rennet. It is an enzyme that catalyzes the coagulation of milk found in the gastric juice of the fourth stomach of young ruminants, and is used in making cheeses and junkets.

repeat unit: It is the simplest structural entity of a polymer chain. So a polymer consists of several repeat units linked together successively along the chain, like the beads of a necklace.

repeats (repeat sequences): It refers to the repeat sequences and approximate repeats that occur throughout the DNA of higher organisms (mammals). For example, the *Alu* sequences of length about 300 characters appear hundreds of thousands of times in human DNA with about 87% homology to a consensus *Alu* string.

repetitive DNA: They are the DNA sequences that are repeated in the genome. These sequences do not code for protein.

repetitive elements: Repetitive elements provide important clues about chromosome dynamics, evolutionary forces, and mechanisms for exchange of genetic information between the organisms The most ubiquitous class of repetitive elements in the DNA sequence in primate genomes is the *Alu* family of interspersed repeats which have arisen in the last 65 million years of evolution.

Alu repeats belong to a class of sequences defined as short interspersed elements (SINEs). Approximately 5,00,000 *Alu* SINEs exist within the human genome, representing about 5% of the genome by mass.

replacement of gene: It is a method of substituting a cloned gene, or part of a gene, which may have been mutated in vitro, for the wild-type copy of the gene.

replacement therapy: It is a therapeutic method of introducing metabolites, co-factors or hormones that are deficient in an organism resulting in a disease.

replica plating: It is a technique in which one or more secondary Petri plates containing different solid (agar-based) selective growth media (lacking nutrients or containing chemical growth inhibitors such as antibiotics) are inoculated with the same colonies of microorganisms from a primary plate (or master dish), reproducing the original spatial pattern of colonies. The technique involves pressing a velvet-covered disk, nitrocellulose membrane, or filter paper to a primary plate, and then imprinting secondary plates with cells in colonies removed from the original plate by the material. Generally, large numbers of colonies (roughly 30-300) are replica plated due to the difficulty in streaking each out individually onto a separate plate.

replication (DNA): It is the basis for biological inheritance. It is a fundamental process occurring in all living organisms to copy their DNA. In this process, each strand of the original double-stranded DNA molecule serves as template for the reproduction of the complementary strand. Hence, following DNA replication, two identical DNA molecules are produced from a single double-stranded DNA molecule.

replication: The repetition of treatments under investigation is known as replication. Replication increases the accuracy of the scope of experiment and enables us to determine the magnitude of the uncontrolled variation that is usually referred to as an error.

replicative form (RF): It is an intermediate form of DNA during the replication of either DNA or RNA viral genomes that is usually double stranded.

replicon: It is a DNA or RNA molecule, or a region of DNA or RNA, that replicates from a single origin of replication.

replisome: It is a complex molecular machine that carries out replication of DNA. It is made up of a number of subcomponents, each providing a specific function during the process of replication.

reporter gene: It is a gene that researchers attach to a regulatory sequence of another gene of interest in cell culture, animals or plants. Certain genes are chosen as reporters because the characteristics they confer on organisms expressing them are easily identified and measured, or because they are selectable markers. Reporter genes are often used as an indication of whether a certain gene has been taken up by or expressed in the cell or organism population.

repressible enzyme: It is an enzyme whose production is generally continuous but can be halted if a particular substance is present in the concentrations greater than normal.

repression: Inhibition of transcription by preventing RNA polymerase from binding to the transcription initiation site: a repressed gene is "turned off."

repressor: It is a DNA-binding protein that regulates the expression of one or more genes by binding to the operator and blocking the attachment of RNA polymerase to the promoter, thus preventing transcription of the genes. This blocking of expression is called repression.

reproduction: It is the biological process by which new "offspring" individual organisms are produced from their "parents". Reproduction is a fundamental feature of all known life; each individual organism exists as a result of reproduction.

reprogenetics: It is a term referring to the merging of reproductive and genetic technologies expected to happen in the near future as techniques like germinal choice technology are available and more powerful. The term was coined by Lee M. Silver, a professor of molecular biology at Princeton University, in his 1997 book *Remaking Eden.*

residue: It refers to a specific monomer within the polymeric chain of a polysaccharide, protein or nucleic acid.

resilience: It refers to the ability of a biotic community to return to its former state after a disturbance.

resistance (drug): It is the reduction in effectiveness of a drug such as an antimicrobial or an antineoplastic in curing a disease or condition. When the drug is not intended to kill or inhibit a pathogen, the term is equivalent to dosage failure or drug tolerance.

resistance factor (R factor): Any of the conjugative or nonconjugative plasmids that carry and transmit genes responsible for resistance to antibiotics from one bacterium to another.

respiration: It refers to the conveyance of oxygen to cells and tissues as carbon dioxide and water are given off. In cells, the conversion of food (e.g., glucose) into energy.

rest period: It is an endogenous physiological condition of viable seeds, buds or bulbs that prevents growth even in the presence of otherwise favourable environmental conditions.

restitution nucleus: It is a nucleus with unreduced or doubled chromosome number that results from the failure of cell division.

restriction endonuclease enzymes: These are the enzymes that cut double-stranded or single stranded DNA at specific recognition nucleotide sequences known as restriction sites. Such enzymes, found in bacteria and archaea, are thought to have evolved to provide a defense mechanism against invading viruses.

restriction exonuclease enzymes: These are the enzymes that work by cleaving nucleotides one at a time from the end of a polynucleotide chain. A hydrolyzing reaction that breaks phosphodiester bonds at either the 3′ or the 5′ end occurs.

restriction fragment length polymorphism (RFLP): It refers to a difference between two or more samples of homologous DNA molecules arising from differing locations of restriction sites, and to a related laboratory technique by which these segments can be distinguished.

restriction fragment: It is a DNA fragment resulting from the cutting of a DNA strand by a restriction enzyme (restriction endonucleases), a process called restriction. Each restriction enzyme is highly specific, recognising a particular short DNA sequence, or restriction site, and cutting both DNA strands at specific points within this site. Most restriction sites are palindromic, (the sequence of nucleotides is same on both strands when read in the 5' to 3' direction), and are four to eight nucleotides long. Many cuts are made by one restriction enzyme because of the chance repetition of these sequences in a long DNA molecule, yielding a set of restriction fragments.

restriction map: It is the map of known restriction sites within a sequence of DNA. Restriction mapping requires the use of restriction enzymes. In molecular biology, restriction maps are used as a reference to engineer plasmids or other relatively short pieces of DNA, and sometimes for longer genomic DNA.

restriction recognition site: It is a location on a DNA molecule containing specific sequences of nucleotides, which are recognized by restriction enzymes. These are generally palindromic sequences and a particular restriction enzyme may cut the sequence between two nucleotides within its recognition site, or somewhere nearby.

retroelement: It refers to any of the integrated retroviruses or the transposable elements that resemble them.

retrogressive succession: It refers to a plant community which, gradually undergoes simplification, a thinning of variety, and a reduction in mass.

retroposon (retrotransposon): They are repetitive DNA fragments which are inserted into chromosomes after they had been reverse transcribed from any RNA molecule. In contrast to retrotransposons, they never encode Reverse Transcriptase (RT). Therefore, they are non-autonomous elements with regard to transposition activity.

retroviral vectors: They are a tool commonly used by molecular biologists to deliver genetic material into cells. This process can be performed inside a living organism (in vivo) or in cell culture (in vitro). Viruses have evolved specialized molecular mechanisms to efficiently transport their genomes inside the cells they infect. Delivery of genes by a virus is termed transduction and the infected cells are described as transduced.

retrovirus: It is a class of eukaryotic RNA viruses. These are RNA viruses that are replicated in a host cell via the enzyme reverse transcriptase to produce DNA from its RNA genome. The DNA is then incorporated into the host's genome by an integrase enzyme. The virus thereafter replicates as part of the host cell's DNA.

retroviruses: They are an unusual form of virus capable of infecting a wide range of vertebrates and causing a number of diseases such as pneumonia, leukemia, and AIDS. Retroviruses have fundamental properties that other virus families do not have in that they are RNA viruses that normally convert their nucleic acid to DNA and then insert this DNA into the genome of the host cell. This provides them with two modes of transmission. They can be both infectious agents that spread horizontally (within a generation) and normal genetic constituents that spread in a Mendelian fashion (vertically between generations). Viruses are important to the study of molecular and cellular biology because they provide simple systems that can be used to manipulate and investigate the functions of many cell types. Numerous studies have demonstrated the utility of animal viruses as probes for investigating different activities of eukaryotic cells. Other examples in which animal viruses have provided important models for biological research of their host cells include studies of DNA replication, transcription, RNA processing, and protein transport.

reversal transfer: It is a special method used in plant tissue culture and involves transfer of a culture from a callus-supporting medium to a shoot-inducing medium.

reverse genetics: It is an approach to discovering the function of a gene by analyzing the phenotypic effects of specific gene sequences obtained by DNA sequencing. This investigative process proceeds in the opposite direction of so-called forward genetic screens of classical genetics. Simply put, while forward genetics seeks to find the genetic basis of a phenotype or trait, reverse genetics seeks to find what phenotypes arise as a result of particular genes.

reverse genetics: It refers to the use of protein information to elucidate the genetic sequence encoding that protein. Used to describe the process of gene isolation starting with a panel of afflicted patients (see also positional cloning).

reverse mutation (reversion): It is the production by further mutation of a premutation gene from a mutant gene. This reverse mutation restores the ability of the gene to produce a functional protein. Strictly, reversion is the correction of a mutation (i.e., it occurs at the same site); more loosely, the term is applied also to a mutation at another site that masks or suppresses the effect of the first mutation (in fact such organisms are not non-mutant, but double mutants with the same phenotype).

reverse transcriptase (RNA-dependent DNA polymerase): It is a DNA polymerase enzyme that transcribes single-stranded RNA into double-stranded DNA. It also helps in the formation of a double helix DNA once the RNA has been reverse transcribed into a single strand cDNA. Normal transcription involves the synthesis of RNA from DNA, hence, reverse transcription is the reverse of this.

Reverse transcriptase-PCR (RT-PCR): It refers to the procedure in which PCR amplification is carried out on DNA that is first generated by the conversion of mRNA to cDNA using reverse transcriptase.

reverse transcription: It is the synthesis of DNA on a template of RNA, accomplished by reverse transcriptase.

RFLP analysis: It is a process in which the DNA sample is broken into pieces (digested) by restriction enzymes and the resulting *restriction fragments* are separated according to their lengths by gel electrophoresis. Although now largely obsolete, RFLP analysis was the first DNA profiling technique inexpensive enough to see widespread application. In addition to genetic fingerprinting, RFLP was an important tool in genome mapping, localization of genes for genetic disorders, determination of risk for disease, and paternity testing.

Rhizobacteria: They are root-colonizing bacteria that form a symbiotic relationship with many legumes.

rhizosphere: It is the narrow region of soil that is directly influenced by root secretions and associated soil microorganisms.

Ri plasmid: A class of large conjugative plasmid responsible for hairy root disease of certain plants. A segment of Ri plasmid is found in the genome of tumour tissue from plants with hairy root disease.

ribonuclease (RNAse): It is a type of nuclease that catalyzes the degradation of RNA into smaller components.

ribonucleic acid: It is a biologically important type of molecule that consists of a long chain of nucleotide units. Each nucleotide consists of a nitrogenous base, a ribose sugar, and a phosphate.

ribose: It is an organic compound with formula $C_5H_{10}O_5$; specifically, a monosaccharide (simple sugar) with linear form H-(C=O)-$(CHOH)_4$-H, which has all the hydroxyl groups on the same side in the Fischer projection.

ribosomal binding site (RBS): It is a sequence on mRNA that is bound by the ribosome when initiating protein translation. It can be either the 5' cap of a messenger RNA in eukaryotes, a region 6-7 nucleotides upstream of the start codon AUG in prokaryotes (called the Shine-Dalgarno sequence), or an internal ribosome entry site (IRES) in viruses.

ribosomal RNA(rRNA): It is the RNA component of the ribosome, the protein manufacturing machinery of all living cells. Ribosomal RNA provides a mechanism for decoding mRNA into amino acids and interacts with tRNAs during translation by providing peptidyl transferase activity. The tRNAs bring the necessary amino acids corresponding to the appropriate mRNA codon.

ribosomes: They are the components of cells that make proteins from amino acids. One of the central tenets of biology, often referred to as the "central dogma," is that DNA is used to make RNA, which, in turn, is used to make protein. The DNA sequence in genes is copied into a messenger RNA (mRNA).

ribozyme (gene shears): They are the RNA molecule that can catalyse chemical reactions, often cutting other RNAs. Based on this, gene sheer technology has been developed. The "Gene Shears" technology involves firstly decoding a gene in an organism that produces a harmful protein. A hammerhead ribozyme is then specially designed to match to that gene code. An army of these specific ribozymes are sent into the cells of the organism.

ribulose biphosphate (RuBP): It is a five-carbon sugar that is combined with carbon dioxide to form a six-carbon intermediate in the first stage of the dark reaction of photosynthesis.

ribulose: It is a ketopentose and a monosaccharide containing five carbon atoms including a ketone functional group.

Richter scale: It refers to an obsolete logarithmic scale of 1-10 that measures earthquake magnitude (but not its felt intensity). Created in California by Charles Richter.

rinderpest (cattle plague): It is an infectious viral disease of cattle, domestic buffalo, and some species of wildlife. It is commonly referred to as cattle plague or steppe murrain. The disease is characterized by fever, oral erosions, diarrhea, lymphoid necrosis, and high mortality.

R-loop: It is the single-stranded loop section of DNA formed by the association of a section of ssRNA with the other strand of DNA in this region, whereby one DNA strand is displaced as the loop.

RNA (ribonucleic acid): It is an organic acid and a class of nucleic acids characterized by the presence of sugar ribose and the organic base uracil. By contrast, DNA contains deoxyribose and thymine, respectively. Most RNA molecules, including messenger RNA and transfer RNA, act as cellular intermediaries, i.e., they convert the genetic information stored in DNA into the proteins that provide cells with structure and enable them to carry out metabolism. In some viruses, RNA also serves as the hereditary material.

RNA editing: It involves molecular processes in which the information content in an RNA molecule is altered through a chemical change in the base makeup. To date, such changes have been observed in tRNA, rRNA, and mRNA molecules of eukaryotes, but not prokaryotes. The demonstration of RNA editing in prokaryotes may only be a matter of time, considering the range of species in which the various RNA editing processes have been found.

RNA polymerase (RNAP or **RNApol):** It is an enzyme that produces RNA. In cells, RNAP is needed for constructing RNA chains from DNA genes as templates, a process called transcription. RNA polymerase enzymes are essential to life and are found in all organisms and many viruses. In chemical terms, RNAP is a nucleotidyl transferase that polymerizes ribonucleotides at the 3' end of an RNA transcript.

RNAase (ribonuclease): It is a type of nuclease that catalyzes the degradation of RNA into smaller components. Ribonucleases can be divided into endoribonucleases and exoribonucleases.

Roentgen(r): It is a unit of measurement for ionizing radiation, named after Wilhelm Conrad Röntgen.

root culture: It is a method in plant tissue culture where isolated root tips of apical or lateral origin are cultured in vitro. Root culture was among the first kinds of plant

root nodule: They occur on the roots of plants (primarily Fabaceae) that associate with symbiotic nitrogen-fixing bacteria. Under nitrogen-limiting conditions, capable plants form a symbiotic relationship with a host-specific strain of bacteria known as rhizobia. This process has evolved multiple times within the Fabaceae.

rotary shaker: It is a rotating apparatus with a platform on which containers can be shaken, such as Erlenmeyer flasks containing cells in liquid nutrient medium.

round worm (*Caenorhabditis elegans*): It is a small, free-living soil nematode, which is the best characterized multicellular animal at the level of genomics, genetics, embryology, and cell and neurobiology, and its genome is fully sequenced. *C. elegans* is unique among model organisms in that it can be grown and genetically manipulated with the speed and ease of a microorganism while offering the features of a real animal. *C. elegans* has a full set of organ systems, complex sensory systems, shows coordinated behaviour, and is further distinguished by the fact that it is possible to trace the lineage of every one of its approximately 1000 constituent cells. Furthermore, the morphology, development, and function of each of its cells have been.

rRNA (ribosomal RNA): It is the RNA component of the ribosome, the protein manufacturing machinery of all living cells. Ribosomal RNA provides a mechanism for decoding mRNA into amino acids and interacts with tRNAs during translation by providing peptidyl transferase activity.

S

S phase (synthesis phase): It is a period in the cell cycle during interphase, between G_1 phase and the G_2 phase. Following G_1, the cell enters the S stage, when DNA synthesis or replication occurs.

saccharifaction: It is the process of breaking a complex carbohydrate (starch or cellulose) into its monosaccharide components.

salinization: It refers to the accumulation of salt on topsoil caused by irrigation. Around the world, 50-75 million acres of once-arable land are too salty to grow food in. Heavily salinized land is not arable until the salt is leached out of it. Several civilizations have fallen because of the resulting ecological damage.

***Salmonella*:** It is a diarrhea-producing intestinal bacterium that causes typhus and poisons raw milk, eggs, meat, and poultry.

Salmonella: It is a genus of rod-shaped, Gram-negative, non-spore forming, predominantly motile enterobacteria. They are a common cause of food poisoning.

salt tolerance (saline resistance): The ability of some plants to withstand a concentration of sodium.

sample: Part of the population selected for study.

sampling error: Errors that arise due to drawing inferences about the population on the basis of a sample are termed sampling errors.

sap (cell): The liquid contained within the vacuole of a plant cell.

saprophyte: An organism, especially a fungus or bacterium, that grows on and derives its nourishment from dead or decaying organic matter.

satellite DNA: It is a part of the chromosome and consists of highly repetitive DNA, and is so-called because repetitions of a short DNA sequence tend to produce a different frequency of the nucleotides, adenine, cytosine, guanine and thymine, and thus have a different density from bulk DNA, such that they form a second or 'satellite' band when genomic DNA is separated on a density gradient.

satellite RNA: It is small, self-splicing RNA molecule that accompanies several plant viruses, including tobacco ring spot virus.

SCA (single chain antigen): Antibody-binding domains in which the two chains are produced by a gene and linked by a short peptide.

Scaffold (chromosomal): It is the central core structure of condensed chromosomes. The scaffold is composed of non-histone chromosomal proteins.

scale up: It refers to the conversion of a process, such as fermentation of a microorganism, from a small scale to a larger scale.

scanning electron microscope (SEM): It is a type of electron microscope that images the sample surface by scanning it with a high-energy beam of electrons in a raster scan pattern. The electrons interact with the atoms that make up the sample producing signals that contain information about the sample's surface topography, composition and other properties such as electrical conductivity.

scarification: It is a natural process important for the germination of many species' seeds. This involves the breaching of the natural seed coating by mechanical, thermal or microbial methods. While this process occurs naturally in the wild, humans have developed techniques to emulate the natural processes for some seeds used in cultivation.

scatter diagram: The scatter of various points in the form of dots

scrapie: It is a fatal, degenerative disease that affects the nervous systems of sheep and goats. It is one of the several transmissible spongiform encephalopathies (TSEs), which are related to bovine spongiform encephalopathy (BSE or "mad cow disease") and chronic wasting disease of deer. It is caused by prions.

secondary antibody: It is an antibody that binds to primary antibodies or antibody fragments. They are typically labelled with probes that make them useful for detection, purification or cell sorting applications. Secondary antibodies may be polyclonal or monoclonal, and are available with specificity for whole Ig molecules or antibody fragments such as the Fc or Fab regions.

secondary immune response: It is the rapid immune response that occurs during the second encounter of the immune system of a mammal with a specific antigen.

secondary infectious disease: It an opportunistic infection imposed on a host with reduced health and resistance resulting from a preceding infection by a more virulent organism.

secondary messenger: They are molecules that relay signals from receptors on the cell surface to target molecules inside the cell, cytoplasm or nucleus. They relay the signals of hormones like epinephrine (adrenalin), growth factors, and others, and cause some kind of change in the activity of the cell.

secondary metabolites: They are organic compounds that are not directly involved in the normal growth, development, or reproduction of organisms. Unlike primary metabolites, absence of secondary metabolites does not result in immediate death, but rather in long-term impairment of the organism's survivability, fecundity, or aesthetics, or perhaps in no significant change at all.

secondary structure (protein): It refers to the organization of peptide backbone of a protein that occurs as a result of hydrogen bonds, e.g., alpha helix, beta pleated sheet.

secretion: It is the process of elaborating, releasing, and oozing chemicals, or a secreted chemical substance from a cell or gland. In contrast to excretion, the substance may have a certain function, rather than being a waste product.

seed drill: It is a sowing device that precisely positions seeds in the soil and then covers them. Before the introduction of the seed drill, the common practice was to "broadcast" seeds by hand. Besides being wasteful, broadcasting was very imprecise and led to a poor distribution of seeds, leading to low productivity. The use of a seed drill can improve the ratio.

seed: It is the matured ovule without accessory parts produced by reproduction.

segmentation polarity gene: It is a gene which defines the anterior and posterior polarities within each embryonic parasegment.

segregant: It is a hybrid resulting from the crossing of two genetically unlike individuals and also differing from both the parents as a result of segregation.

segregation: It is the process of separation of allelic genes that occurs typically during meiosis. The result is seen as the separation of alleles from each other in the gametes of heterozygotes; the occurrence of different phenotypes among offspring, resulting from chromosome or allele separation in their heterozygous parents.

selectable marker: It is a gene introduced into a cell, especially a bacterium or to cells in culture, that confers a trait suitable for artificial selection. They are a type of reporter gene used in microbiology, molecular biology, and genetic engineering to indicate the success of a transfection or other procedure meant to introduce foreign DNA into a cell. Selectable markers are often antibiotic resistance genes; bacteria that have been subjected to a procedure to introduce foreign DNA are grown on a medium containing an antibiotic, and the bacterial colonies that can grow have successfully taken up and expressed the introduced genetic material.

selection coefficient: It is a measure of disadvantage of a given gene or mutation that is usually the frequency of individuals having the gene or mutation, which fail to leave fertile offspring.

selection culture: It is a selection based on differences in environmental conditions or culture medium composition, such that preferred variant cells or cell lines are favoured over other variants or the wildtype.

selection pressure: It is the pressure caused by those factors that influence the direction of natural selection. It acts on the selection acting on a population of organisms or cells in culture.

selection response: It is the difference between the mean of the individuals selected to be parents and the mean of their offspring.

selection unit: It is a biological entity within the hierarchy of biological organisation that is subject to natural selection.

selection: It is a procedure for either isolating or identifying specific organisms in a mixed culture.

selective agent: It refers to any factor that brings about differences in fertility and mortality.

self-fertilization: The process by which pollen of a given plant fertilizes the ovules of the same plant.

self-incompatibility: It is a general name for several genetic mechanisms in angiosperms, which prevent self-fertilization and thus encourage out-crossing.

self-pollination: It is type of pollination that can occur when a flower has both stamen and a carpel(pistil) in which the cultivar or species is self fertile and the stamens and the sticky stigma of the carpel contact each other in order to accomplish pollination.

self-replicating elements: They are the extrachromosomal DNA elements like plasmids that have origins of replication for the initiation of their own DNA synthesis.

self-replicating machine: It is an artificial construct that is theoretically capable of autonomously manufacturing a copy of itself using raw materials taken from its environment self-sterility.

self-replication: It is the process by which some things make copies of themselves. Biological cells, given suitable environments, reproduce by cell division. Biological viruses can reproduce, but only by commandeering the reproductive machinery of cells through a process of infection.

SEM (scanning electron microscope): It is a type of electron microscope that images the sample surface by scanning it with a high-energy beam of electrons in a raster scan pattern. The electrons interact with the atoms that make up the sample producing signals that contain information about the sample's surface topography, composition and other properties such as electrical conductivity.

semen sexing or sperm sexing or sperm sorting technology: It is a technique used to predetermine the sex of offspring prior to conception, thereby maximizing productivity, profitability, and genetic potential.

semi-conservative replication: It is the type of replication that occurs in nature. During DNA duplication, the parent DNA template would produce two copies, each containing one of the original strands and one new strand. The semi-conservative theory can be confirmed by making use of the fact that DNA is made up of nitrogen bases.

semi-continuous culture: It is a type of tissue culture in which the cells in an actively dividing state are maintained in culture by periodically draining off the medium and replenishing it with fresh medium.

semi-liquid or semi-solid: The term is usually used in case of a nutrient medium and refers to a state called gelled but not firmly so. Small amounts of a gelling agent are used to obtain a semi-solid medium, also called semi-liquid.

semi-permeable membrane or selectively-permeable membrane or **partially-permeable membrane** or **differentially-permeable membrane**: It is a membrane that allows certain molecules or ions to pass through it by diffusion and occasionally specialized "facilitated diffusion".

semi-sterility or half sterility: The phenotype of individuals heterozygotic for certain types of chromosome aberration; expressed as a reduced number of viable gametes and hence reduced fertility. Nonviability of a proportion of gametes or zygotes.

senescence: Also called biological aging, it is the change in the biology of an organism as it ages after its maturity. Such changes range from those affecting its cells and their function to that of the whole organism. There are a number of theories as to why senescence occurs, including ones that claim it is programmed by gene expression changes and is the accumulative damage of biological processes.

sense RNA: It is an RNA molecule produced by transcription that contains a coding.

sensitivity: It is a detectable change in the internal or external environment. The ability of an organism or organ to respond to external stimuli is called sensitivity.

sepsis: It is a serious medical condition that is characterized by a whole-body inflammatory state and the presence of a known or suspected infection.

sequence hypothesis: It is a theory which was first formally proposed in a review on Protein Synthesis by Francis Crick in 1958. It states that the sequence of bases in the genetic material (DNA or RNA) determines the sequence of amino acids for which that segment of nucleic acid codes, and this amino acid sequence determines the three-dimensional structure into which the protein folds.

Sequence Tagged Site (STS): It refers to the unique sequence from a known chromosomal location that can be amplified by PCR. STSs act as physical markers for genomic mapping and cloning.

sequence-tagged site (STS): It is a relatively short, easily PCR-amplified sequence (200 to 500 bp) which can be specifically amplified by PCR and detected in the presence of all other genomic sequences and whose location in the genome is mapped.

sequencing: It is a method of determining the primary structure (or primary sequence) of an unbranched biopolymer. Sequencing results in a symbolic linear depiction known as a **sequence** which succinctly summarizes much of the atomic-level structure of the sequenced molecule.

serial float culture: It is a method of haploid culture and is a technique of floating anthers on liquid medium developed by Sunderland. Anther dehiscence, pollen release and development occur at intervals of several days, and in different nutrient media.

serology: It is the scientific study of blood serum. In practice, the term usually refers to the diagnostic identification of antibodies in the serum. Such antibodies are typically formed in response to an infection against other foreign proteins (in response, for example, to a mismatched blood transfusion), or to one's own proteins.

serum albumin: It is a globular protein obtained from blood and body fluids. It is the most abundant plasma protein in mammals. Albumin is essential for maintaining the osmotic pressure needed for proper distribution of body fluids between intravascular compartments and body tissues.

sewage treatment: It is one of the most widespread biotechnological processes. It is the process of removing contaminants from wastewater and household sewage, both run-off and domestic. It includes physical, chemical, and biological processes to remove physical, chemical and biological contaminants. Its objective is to produce a waste stream (or treated effluent) and a solid waste or sludge suitable for discharge or reuse back into the environment.

sex chromosomes (allosomes): These are the chromosomes connected with the determination of sex. The nuclei of human cells contain 22 autosomes and 2 sex chromosomes.

sex determination: It is a biological system that determines the development of sexual characteristics in an organism. Most sexual organisms have two sexes. In many cases, sex determination is genetic: males and females have different alleles or even different genes that specify their sexual morphology. In animals, this is often accompanied by chromosomal differences. In other cases, sex is determined by environmental variables (such as temperature) or social variables (the size of an organism relative to other members of its population). The details of some sex-determination systems are not yet fully understood.

sex factor or fertility factor or F factor, sex factor or F-plasmid: It is a bacterial DNA sequence that allows a bacterium to produce a sex pilus necessary for conjugation.

sex hormones: They are hormones, such as estrogen and androgen, affecting the growth or function of the reproductive organs, the development of secondary sex characteristics, and the behavioural patterns of animals.

sex linkage: It is the phenotypic expression of an allele related to the chromosomal sex of the individual.

sex mosaic (gynandromorph): It refers to an organism having both male and female characteristics, especially an insect exhibiting a mixture of male and female tissues or sex organs.

sexduction: It is the incorporation of bacterial genes into F factors and their subsequent transfer, by conjugation, to a recipient cell.

sexed embryos: Embryos separated according to sex.

sex-limited genes: They are genes which are present in both sexes of sexually reproducing species but turned on in only one sex. In other words, sex-limited genes cause the two sexes to show different traits or phenotypes. An example of sex-limited genes are genes which instruct male elephant seal to grow big and fight, at the same time instructing female seals to grow small and avoid fights. These genes are responsible for sexual dimorphism.

sex-linked dominant: It is a rare way that a trait or disorder can be passed down through families. A single abnormal gene on the X chromosome can cause a sex-linked dominant disease.

Sexual PCR (Molecular Diversity): It refers to a form of PCR in which similar, but not identical, DNA sequences are reassembled to obtain novel juxtapositions, simulating the result of genetic recombination. The result is the creation of an array of related genes which may possess improved characteristics. By repeated rounds of recombination, selection and PCR-based amplification vastly improved gene products, such as enzymes with greater activity, may be generated and selected.

sexual reproduction: It is the process where two gametes fuse to form one fertilized cell or zygote.

shake culture: It is a method for isolating anaerobic bacteria by shaking a deep liquid culture of an agar or gelatin to distribute the inoculum before solidification of the medium. A liquid medium in a flask that has been inoculated with an aerobic microorganism is placed on a shaking machine; action of the machine continually aerates the culture.

shakers: They are the equipment used in the culture of cells or tissues. The shaking speed can be adjusted for gentle and even agitation of suspension cultures. There are different types of shakers used in plant tissue culture.

Shine-Dalgarno sequence: It was proposed by Australian scientists John Shine and Lynn Dalgarno, and is a ribosomal binding site in the mRNA, generally located 8 basepairs upstream of the start codon AUG.

shoaling: It refers to a group of fish that stay together for social reasons and tends to swim in the same direction in a coordinated manner.

shoot-tip grafting (micrografting): It is a method in which a shoot tip or meristem tip is grafted onto a prepared seedling or micropropagated rootstock in culture. Meristem tip grafting is mainly used for virus elimination in plants.

short interspersed nuclear elements (SINEs): They are a family of short, moderately repetitive elements of eukaryotes, occurring about 1,00,000 times in a genome. SINES appear to be DNA copies of certain tRNA molecules, created presumably by the unintended action of reverse transcriptase during retroviral infection.

short template: A DNA strand that is synthesized during the polymerase chain reaction.

short-day plants: They are the plants that flower only after being exposed to light periods shorter than a certain critical length, as in early spring or fall, e.g., Chrysanthemums, strawberries etc.

shotgun cloning: It refers to the cloning of an entire gene segment or genome by generating a random set of fragments using restriction endonucleases to create a gene library that can be subsequently mapped and sequenced to reconstruct the entire genome.

shuttle vector (bifunctional vector): It is a type of cloning vector that contains sequences enabling it to be propagated and maintained in more than one type of host, e.g., YEp, pJDB219, etc.

sib: It refers to a blood relative; one of a group of persons all descended from a common ancestor.

siderophore: They are small, high-affinity iron chelating compounds secreted by microorganisms such as bacteria, fungi and grasses. Siderophores are amongst the strongest soluble Fe^{3+} binding agents known.

sieve cell: They are long and slender sieve element with relatively unspecialized sieve areas and with tapering end walls that lack sieve plates. They are associated with phloem in plants.

sieve tube: It is a tube within the phloem tissue of a plant composed of joined sieve elements.

sievert: It is the SI derived unit of dose equivalent. It attempts to reflect the biological effects of radiation as opposed to the physical aspects, which are characterised by the absorbed dose, measured in gray. It is named after Rolf Sievert, a Swedish medical physicist famous for work on radiation dosage measurement and research into the biological effects of radiation.

sigma factor: It is a prokaryotic transcription initiation factor that enables specific binding of RNA polymerase to gene promoters. Different sigma factors are activated in response to different environmental conditions. Every molecule of RNA polymerase contains exactly one sigma factor subunit.

signal peptide or targeting signals or signal sequences or transit peptides or localization signals: It is a short (3-60 amino acids long) peptide chain that directs the transport of a protein.

signal sequence (leader sequence): It refers to a short sequence added to the amino-terminal end of a polypeptide chain that forms an amphipathic helix allowing the nascent polypeptide to migrate through membranes such as the endoplasmic reticulum or the cell membrane. It is cleaved from the polypeptide after the protein has crossed the membrane.

signal sequence: It is a peptide present on proteins that is destined either to be secreted or to be membrane components. It is usually at the N terminus and normally absent from the mature protein.

signal transduction: The biochemical events that conduct the signals in the living organisms. It is a mechanism that converts a mechanical/chemical stimulus to a cell into a specific cellular response. Signal transduction starts with a signal to a receptor, and ends with a change in cell function.

signal-to-noise ratio (SNR or S/N): It is a measure used in science and engineering to quantify how much a signal has been corrupted by noise. It is defined as the ratio of signal power to the noise power corrupting the signal. A ratio higher than 1:1 indicates more signal than noise. While SNR is commonly quoted for electrical signals, it can be applied to any form of signal.

significant level: The statistical tests fix the probability of committing Type I error at a certain level and minimize the chances of committing Type II error.

silencer (DNA): It is a DNA sequence capable of binding transcription regulation factors termed repressors. Upon binding, RNA polymerase is prevented from initiating transcription, thus decreasing or fully suppressing RNA synthesis.

silica: It is a transparent mineral of silicon dioxide found in glass, quartz, opals.

similarity (homology) search: It refers to a newly sequenced gene. There are two main approaches to the prediction of structure and function from the amino acid sequence. Homology methods are the most powerful and are based on the detection of significant extended sequence similarity to a protein of known structure, or of a sequence pattern characteristic of a protein family. Statistical methods are less successful but more general and are based on the derivation of structural preference values for single residues, pairs of residues, short oligopeptides or short sequence patterns. The transfer of structure/function information to a potentially homologous protein is straightforward when the sequence similarity is high and extended in length, but the assessment of the structural significance of sequence similarity can be difficult when sequence similarity is weak or restricted to a short region.

simple sequence repeat (SSR) or microsatellite: They are repeating sequences of 1-6 base pairs of DNA and are typically neutral and co-dominant. They are used as molecular markers in genetics, for kinship, population and other studies. They can also be used to study gene duplication or deletion.

simulated annealing (SA): It is a generic probabilistic metaheuristic for the global optimization problem of applied mathematics, namely locating a good approximation to the global optimum of a given function in a large search space. It is often used when the search space is discrete (e.g., all tours that visit a given set of cities). For certain problems, simulated annealing may be more effective than exhaustive enumeration, provided that the goal is merely to find an acceptably good solution in a fixed amount of time, rather than the best possible solution.

single copy DNA: It is copy of DNA which contains the structural genes (protein-coding sequences.

Single nucleotide polymorphisms (SNPs): It refers to the variations that are seen in single base pairs scattered throughout the human genome that serve as measures of genetic diversity in humans. About 1 million SNPs are estimated to be present in the human genome, and SNPs are useful markers for gene mapping studies.

single-cell line (cell strain): Cell culture is the complex process by which cells are grown under controlled conditions. Cell strain is a type of cell culture initiated from a single cell, usually from suspension cultures of single cells or small aggregates plated on solidified medium. The latter may incorporate a selective agent, from which tolerant or resistant individual cell lines or cell clones can be selected.

single-cell protein (SCP): It typically refers to sources of mixed protein extracted from pure or mixed cultures of algae, yeasts, fungi or bacteria (grown on agricultural wastes) used as a substitute for protein-rich foods, in human and animal feeds.

single-chain antigen (SCA): Antibody-binding domain in which the two chains are produced by a gene and linked by a short peptide.

single-node culture: It is a culture of separate lateral buds, each carrying a piece of stem tissue.

single-nucleotide polymorphism (SNP): It is a DNA sequence variation occurring when a single nucleotide — A, T, C, or G — in the genome (or other shared sequence) differs between members of a species or paired chromosomes in an individual. For example, two sequenced DNA fragments from different individuals, AAGCCTA to AAGCTTA, contain a difference in a single nucleotide.

single-pass sequencing: It refers to the rapid sequencing of large segments of the genome of an organism by isolating as many expressed (cDNA) sequences as possible and performing single sequencer runs on their 5′ or 3′ ends. Single-pass sequencing typically results in individual, error-prone sequencing reads of 400-700 bases, depending on the type of sequencer used. However, if many of these are generated from numerous clones from different tissues, they may be overlapped and assembled to remove the errors and generate a contiguous sequence for the entire expressed gene.

single-strand-DNA-binding protein (SSB or SSBP): It binds single-stranded regions of DNA to prevent premature annealing. The strands have a natural tendency to revert to the duplex form, but SSB binds to the single strands, keeping them separate and allowing the DNA replication machinery to perform its function.

single-stranded: A term used to describe nucleic acid molecules consisting of only one polynucleotide chain.

site: It refers to the sites in sequences that can be located either in DNA (e.g. binding sites, cleavage sites) or proteins. In order to identify a site in DNA, ambiguity symbols are used to allow several different symbols at one position. Proteins, however, need a different mechanism (see Pattern). Restriction enzyme cleavage sites, for instance, have the following properties: limited length (typically, less than 20 base pairs); definition of the cleavage site and its appearance (3', 5' overhang or blunt); definition of the binding site.

site-directed mutagenesis (site-specific mutagenesis or oligonucleotide-directed mutagenesis): It is a molecular biology technique in which a mutation is created at a defined site in a DNA molecule, typically a circular molecule known as a plasmid. In general, this form of mutagenesis requires that the wild type gene sequence be known.

site-specific: A term used to describe any process or enzyme which acts at a defined sequence within a DNA or RNA molecule.

six-base cutter: It is a type II restriction endonuclease that binds (and subsequently cleaves) DNA at sites that contain a sequence of six nucleotide pairs uniquely recognized by that enzyme.

skewness: The extent to which a distribution of data points is concentrated at one end or the other; the lack of symmetry.

small nuclear ribonucleoprotein (snRNP): They are RNA-protein complexes that combine with unmodified pre-mRNA and various other proteins to form a spliceosome, a large RNA-protein molecular complex upon which splicing of pre-mRNA occurs. The action of snRNPs is essential for the removal of introns from pre-mRNA, a critical aspect of post-transcriptional modification of RNA, occurring only in the nucleus of eukaryotic cells.

small nuclear RNA (snRNA): It is a class of small RNA molecules found within the nucleus of eukaryotic cells. They are transcribed by RNA polymerase II or RNA polymerase III and are involved in a variety of important processes such as RNA splicing (removal of introns from hnRNA), regulation of transcription factors (7SK RNA) or RNA polymerase II (B2 RNA), and maintaining the telomeres. They are always associated with specific proteins, and the complexes are referred to as small nuclear ribonucleoproteins (snRNP) or sometimes as snurps.

soil amelioration: It is a process of improvement of poor soils, usually using bacteria or fungi.

soil: It is a mixture of humus and particles weathered from rock. A cubic yard of soil will cover 300 square feet of ground to a depth of 1 inch.

soilless culture (soil-free culture): It is a method of growing plants in nutrient solution or medium without soil.

solid media: Type of nutrient media used in bacterial, plant and animal tissue culture that has been prepared by adding agar or some other similar substance, mostly inert solidifying agent.

solvent accessibility: It refers to the surface area (typically measured in square angstroms) of a biological molecule, usually a protein, that is exposed to solvent in its native, folded form. Determining the solvent accessibility of a protein helps define which amino acids in its molecular sequence are on the exterior of the molecule, and thus available to participate in interactions with other molecules.

somaclonal variation: It is a term used to describe the variation seen in plants that have been produced by plant tissue culture. Chromosomal rearrangements are an important source of this variation.

somatic (embryoid): It refers to an organized embryonic structure morphologically similar to a zygotic embryo but initiated from non-zygotic cells, which is in vitro in origin.

somatic (parasexual) hybridization: It is the hybridization by induced fusion of cells (protoplasts) from two contrasting genotypes for production of somatic hybrids called cybrids, which contain various mixtures of nuclear and cytoplasmic genomes.

somatic cell embryogenesis: The process in which embryos are produced either from somatic cells of explants (direct embryogenesis) or by induction on callus formed by explants (indirect embryogenesis).

somatic cell gene therapy: It is a technique in which the therapeutic genes are transferred into the somatic cells of a patient. Any modifications and effects will be restricted to the individual patient only, and will not be inherited by the patient's offspring or later generations.

somatic cell nuclear transfer (SCNT): It is a laboratory technique for creating a clonal embryo, using an ovum with a donor nucleus. It can be used in embryonic stem cell research, or, potentially, in regenerative medicine where it is sometimes referred to as "therapeutic cloning." It can also be used as the first step in the process of reproductive cloning.

somatic cell variant: It is a somatic cell with unique characters not present in other cells, such as might be selected for in a screening trial that follows a mutation event.

somatic cells: They are cells forming the body of an organism, as opposed to reproductive or germline cells.

somatic hypermutation (SHM): It is a mechanism inside cells that is part of the way the immune system adapts to the new foreign elements that confront it. SHM diversifies the receptors used by the immune system to recognize foreign elements (antigens) and allows the immune system to adapt its response to new threats during the lifetime of an organism.

somatic reduction: It is a process of halving of the chromosomal number of somatic cells and is a possible method of producing haploids from somatic cells and calluses by artificial means. Recurrent somatic reduction is a normal ontogenetic process in apogeotropic roots of cycads, which develop into dichotomously branching coralloid masses.

somatocrinin: It is a hormone, produced in the hypothalamus, which promotes the secretion of growth hormones; a growth-hormone-releasing hormone (GHRH).

somatostatin (growth hormone-inhibiting hormone or somatotropin release-inhibiting factor): It is a peptide hormone that regulates the endocrine system and affects neurotransmission and cell proliferation via interaction with G-protein-coupled somatostatin receptors and inhibition of the release of numerous secondary hormones.

somatotrophin (somatotropin): A growth hormone produced by the anterior pituitary gland. It promotes growth in humans.

sonication: It is the technique of applying sound (usually ultrasound) energy to agitate particles in a sample for various purposes. In the laboratory, it is usually applied using an ultrasonic bath or an ultrasonic probe, colloquially known as a sonicator. In a paper machine, an ultrasonic foil can distribute cellulose fibres more uniformly and strengthen the paper.

SOS response: Post-replication DNA repair system that allows DNA replication to bypass lesions or errors in the DNA is called SOS response. The SOS uses the RecA protein (Rad51 in eukaryotes). The RecA protein, stimulated by single-stranded DNA, is involved in the inactivation of LexA repressor thereby inducing the response. It is an error-prone repair system.

source DNA: It is the DNA from any organism that contains a particular target gene, which is used as a starting material in a cloning experiment.

source organism: It refers to an organism from which DNA is purified and used in a cloning experiment.

Southern blot: A method routinely used for detection of a specific DNA sequence in DNA samples. Southern blotting combines transfer of electrophoresis-separated DNA fragments to a filter membrane and subsequent fragment detection by probe hybridization. The method is named after its inventor Edwin Southern.

Southern hybridization: A procedure in which a cloned, labelled segment of DNA is hybridized to DNA restriction fragments on a Southern blot. In Southern hybridization, a sample of DNA containing fragments of different sizes is subjected to electrophoresis using either polyacrylamide or agarose gel.

sparger: It is a device used to introduce air or gas into a bioreactor air in the form of separate fine streams.

speciation: It is the evolutionary process by which a new biological species arises. Orator F. Cook coined this term.

species: It refers to a group of interbreeding organisms which are morphologically and genetically identical.

spent medium: It is the medium that needs to be discarded because it has been depleted of nutrients, dehydrated or accumulated toxic metabolic products.

sperm (spermatozoon): It is the haploid cell, i.e., the male gamete. It joins an ovum to form a zygote. A zygote is a single cell, with a complete set of chromosomes, that normally develops into an embryo.

sperm competition: It refers to the competition between different spermatozoa to reach and fertilize the single egg cell of a single female.

spermatid: It refers to one of the four cells formed by the meiotic divisions in spermatogenesis.

spermatocyte: It refers to the sperm mother cell that undergoes two meiotic divisions (spermatogenesis) to form four spermatids; the *primary* spermatocyte before completion of the first meiotic division; the *secondary* spermatocyte after completion of the first meiotic division.

spermatogenesis: It is a series of cell divisions in the testis by which maturation of the gametes (sperm) of the male takes place.

spermatogonium: The primordial male germ cell that may divide by mitosis to produce more spermatogonia.

spirillum: It refers to a bacterium with a cell body that twists like a spiral. *Spirillum* is also a genus of Gram-negative bacteria.

spirochaete: These are distinctive non-rigid, corkscrew-shaped Gram-negative bacteria, which have long, helically coiled cells and move by means of muscular flexions of the cell.

splice form: By using alternative splicing, a single message precursor from DNA can generate an entire family of mRNAs and proteins. This can be utilized to create specificity in cell-cell or cell-ligand interactions. A cell may produce a given protein, but it will be a different splice-form of the protein than that produced by an adjacent cell. In this manner, the two cells have the potential to interact differently with other cells or molecules. Two places where this has been extremely important is in the production of cell-surface specificity proteins in the immune and nervous systems.

splice site: The sequence found at the 5′ and 3′ region of exon/intron boundaries, usually defined by a consensus sequence.

spliceosomes: It is a complex of specialized RNA and protein subunits that remove introns from a transcribed pre-mRNA (hnRNA) segment.

splicing: It refers to the joining together of separate DNA or RNA component parts. For example, RNA splicing in eukaryotes involves the removal of introns and the stitching together of exons from the pre-mRNA transcript before maturation.

split genes: In higher organisms, unlike in prokaryotes the structural genes are typically divided into a number of non-coding regions called introns and coding regions called exons. Hence, the genes here are called split genes.

spore: It is a reproductive cell that develops usually by asexual methods into an individual without any fusion with other cells.

sporocyte: It refers to a diploid cell that gives rise to four haploid spores by meiosis.

sporophyll: It is a modified leaf producing sporangia in pteridophytes and gymnosperms.

sporophyte: The diploid generation in the life cycle of a plant is called a sporophyte. It is a spore-bearing (asexual) generation in plants that exhibit alternation of generations.

sport: The term was introduced by Charles Darwin to indicate an individual or portion thereof distinguished by a spontaneous mutation.

SSR (simple sequence repeats): They are short DNA sequences made up of nucleotides that are repeated in tandem, like Di-nucleotide repeat: CACACACA; Tri-nucleotide repeat: ATGATGATGATG. It is believed that when DNA is being replicated, errors occur in the process and extra sets of these repeated sequences are added to the strand. Over time, these repeated sequences vary in length between one cultivar and another.

staggered cuts: They are the cleavage of two opposite strands of duplex DNA at points near one another.

stamen: The pollen-producing male reproductive organ of a flower, usually consisting of a filament and an anther.

standard deviation: It is a statistical measure of variability and is the square root of its variance. Standard deviation is a widely used measure of the variability or dispersion, being algebraically more tractable though practically less robust than the expected deviation or average absolute deviation.

standard deviation: It is the square root of variance. An absolute measure of dispersion which is expressed in the same units as the original data, rather than in the squared units of the variance.

standard error of estimate: A measure of the reliability of the estimating equation, indicating the variability of the observed points around the regression line, i.e., the extent to which observed values differ from their predicted values on the regression line.

standard error: It is a statistical measure of variation in a population of means, used to indicate how well sample estimates represent population parameters.

standard normal curve: The curve with zero mean and unit standard deviation is known as the standard normal curve.

standing crop: It refers to the biomass of a trophic level, community, population, or species at a certain time and place.

starch (amylum): It is a carbohydrate consisting of a large number of glucose units joined together by glycosidic bonds. This polysaccharide is produced by all green plants as an energy store.

start codon (initiator codon): It is the codon (AUG) on a messenger RNA molecule where protein synthesis begins and is a set of three nucleotides in an mRNA molecule with which the ribosome starts the process of translation. The start codon sets the reading frame for translation.

stationary culture: It is a tissue culture kept in the growth chamber with no agitation movement. The shakers are not used in this case.

stationary phase: It is a culture showing the slow growth rate as a result of nutrient depletion and accumulation of toxic products. This phase is reached as the bacteria begin to exhaust the resources that are available to them. This phase is a constant value as the rate of bacterial growth is equal to the rate of bacterial death.

statistic: It is an estimate based on a sample or samples of a population, providing an indication of the true population parameter.

statistics: A measure computed from the data of a sample.

steady state: It is a situation in which all state variables are constant in spite of ongoing processes that strive to change them. In a continuous fermentation process, the condition when the number of cells that are removed with the outflow is exactly balanced by newly synthesized cells.

stele: The central cylinder of vascular tissues in the stems of vascular plants.

stem cell treatment: It is a type of intervention strategy that introduces new cells into damaged tissue in order to treat disease or injury.

stem cell: An undifferentiated active somatic cell characterized by the ability to renew themselves through mitotic cell division and differentiate into a diverse range of specialized cell types.

sterile: It refers to a condition with no perceptible or viable microorganisms. It also refers to a situation where an organism is incapable of fertilizing or being infertile. The enability to produce offspring is called sterility.

sterilization: The technique of sterilizing the media or equipment.

Steward bottle: It is the flask developed by Steward for the growth of cells and tissues in a liquid medium, in which they can be periodically submerged during rotation.

sticky ends (cohesive ends): The single-stranded nucleotide sequence left on a restriction fragment by type II restriction enzymes that cut each strand at a separate location. These unpaired regions are available for hybridization with complementary ends on other fragments during the creation of recombinant DNA. The two DNAs have single strands protruding from their ends, and if these single strands are able to pair with each other by the same rules that DNA strands are held together, then these two molecules could come together.

stirred-tank fermenter: It is the growth vessel in which cells or microorganisms are mixed by mechanically-driven impellers.

stock plant: It is the source plant from which cuttings or explants are made. Stock plants are usually maintained carefully in an optimum state for explant use.

stock solutions: Pre-prepared solutions of individual components used to prepare many different types of media.

stop codon (termination codon): It is a nucleotide triplet within the messenger RNA that signals the termination of translation. Proteins are based upon polypeptides which are unique sequences of amino acids, and most codons in messenger RNA correspond to the addition of an amino acid to a growing polypeptide chain which may ultimately become a protein — stop codons signal the termination of this process, releasing the amino acid chain.

strain isolation: It is a technique of isolation of any bacterium, animal or plant from the culture.

strain: It is a group of individuals from a common origin within a species.

stratification: The technique of treatment of moist seeds at low temperature (+2° – +4°C) to break dormancy.

stratified random sampling: In stratified random sampling, first the population is divided into relatively homogeneous groups or strata and a random sample is drawn from each group or stratum to produce an overall sample.

stress proteins: It refers to proteins produced in response to environmental stress. Some of them ward off the hypothermia brought by extreme cold.

stringency: It is the quality or state of being stringent. In molecular biology it refers to the conditions like notably temperature, salt concentration(s) and pH – that dictate the annealing of single-stranded DNA/DNA, DNA/RNA and RNA/RNA hybrids. At high stringency, duplexes form only between strands with perfect one-to-one complementarity; lower stringency allows annealing between strands with some degree of mismatch between the bases. A plasmid that only replicates along with the main bacterial chromosome and is present as a single copy, or at most several copies, per cell is called a stringent plasmid.

structural gene: It is a gene that codes for any RNA or protein product other than a regulatory factor (i.e., regulatory protein). It may code for a structural protein, an enzyme, or an RNA molecule not involved in regulation.

structural gene: It refers to the gene which encodes a structural protein.

structure prediction: It refers to the algorithms that predict the secondary, tertiary and sometimes even quarternary structure of proteins from their sequences. Determining protein structure from sequence has been dubbed "the second half of the Genetic Code", since it is the folded tertiary structure of a protein that governs how it functions as a gene product. As yet most structure prediction methods are only partially successful, and typically work best for certain well-defined classes of proteins.

sub-culturing: It is the technique of culturing of tissue or plantlets or bacteria by transferring to a fresh medium. Sometimes used to denote the adding of fresh liquid to a suspension culture.

substitution (point mutation or single base substitution): It is a type of mutation that causes the replacement of a single base nucleotide with another nucleotide of the genetic material, DNA or RNA. Often the term point mutation also includes insertions or deletions of a single base pair.

substitution matrix: It refers to the model of protein evolution at the sequence level resulting in the development of a set of widely used substitution matrices. These are frequently called Dayhoff, MDM (Mutation Data Matrix), BLOSUM or PAM (Percent Accepted Mutation) matrices. They are derived from global alignments of closely related sequences. Matrices for greater evolutionary distances are extrapolated from those for lesser ones.

substrate: It is the compound that is altered or broken down by an enzyme.

subtraction library: It refers to the cDNA library that only contains cDNAs uniquely expressed in a given cell or tissue. For example, T cells and B cells will express many common RNAs, as well as a very small percentage which will be unique for T cells and B cells respectively. To make a T cell subtraction library, the cDNA from a T cell library is hybridized with a vast excess of B cell RNA. The commonly expressed genes will result in RNA-cDNA hybrids, which can be removed (or subtracted) to leave only T cell specific cDNAs.

succession (ecological): It refers to the gradual replacement of one plant community by another. Brought about by changes in climate, environment, or that the community.

sucrose density gradient centrifugation: It is the procedure used to fractionate mRNAs or DNA fragments on the basis of size.

superbug: It is the term used for the bacterial strain of *Pseudomonas* developed by Anand Chakrabarty, who combined hydrocarbon-degrading genes carried on different plasmids into one organism. The term is also used for any bacterium with extraordinary abilities.

supercoil: The two strands twist around the helical axis once every 10.4-10.5 base pairs of sequence. Adding or subtracting twists, as some enzymes can do, imposes the strain. If a DNA segment under twist strain were to be closed into a circle by joining its two ends and then allowed to move freely, the circular DNA would contort into a new shape, such as a simple figure—eight. Such a contortion is a supercoil.

supercoiled plasmid: The predominant in vivo form of plasmids, in which the plasmid is coiled around histone-like proteins. Supporting proteins are stripped away during extraction from the bacterial cell, causing the plasmid molecule to supercoil around itself in vitro.

supergene: It is a group of neighbouring genes on a chromosome which are inherited together because of close genetic linkage and are functionally related in an evolutionary sense, although they are rarely co-regulated genetically.

superhydrophobic: It refers to the surfaces such as the leaves of the lotus plant have highly hydrophobic surfaces, i.e., extremely difficult to wet. The contact angles of a water droplet exceeds 150° and the roll-off angle is less than 10°.

supernatant: It is the soluble liquid fraction of a sample after centrifugation or precipitation of insoluble solids.

suppressor gene: A gene that suppresses the phenotypic expression of another gene.

suppressor mutation: It is a mutation that counteracts the effects of another mutation. A suppressor maps at a different site than the mutation it counteracts, either within the same gene or at a more distant locus.

suppressor-sensitive mutant: It is conditionally lethal, host range, bacteriophage mutant that produces nonsense codons and can replicate only in a host bacterium able to translate the nonsense codon; the mutation's effects are lethal (i.e., prevents the replication of the virus) in a bacterium without such a suppressor mechanism.

surfactant: It a material that can greatly reduce the surface tension of water when used in very low concentrations. A popular surfactant, *Softanol-90* reduces the surface tension of water from 73 to 30 dynes per centimetre when used at a concentration of 0.005 per cent. Ethanol at a concentration of 20 per cent, reduces tension of water to 38 dynes per centimetre.

susceptibility: It is a state or character of being susceptible.

susceptible: It refers to a member of a population who is at risk of becoming infected by a disease, or cannot take a certain medicine, antibiotic, etc., if he or she is exposed to the infectious agent.

suspension culture: It is common for microorganisms but less so for the culture of cells of most multicellular organisms. When referring to mammalian cells, suspension culture is used for the maintenance of cell types, which do not adhere, including some types of blood cells, or in order to have cells express characteristics, which are not seen in the adherent form.

sustainomics: It is an interdisciplinary and interactive branch that deals with comprehensive, balanced and heuristic metaframework of making development sustainable.

swarm intelligence (SI): It describes the collective behaviour of decentralized, self-organized systems, natural or artificial. The concept is employed in work on artificial intelligence. The expression was introduced by Gerardo Beni and Jing Wang in 1989, in the context of cellular robotic systems.

swarm robotics: It is a new approach to the coordination of multirobot systems which consist of large numbers of mostly simple physical robots. It is supposed that a desired collective behaviour emerges from the interactions between the robots and interactions of robots with the environment. This approach emerged on the field of artificial swarm intelligence, as well as the biological studies of insects, ants and other fields in nature, where swarm behaviour occurs.

symbiont: It is an organism in a symbiotic relationship. It is also called symbiote. It lives in close association with another different kind of organism that is beneficial to both.

symbiosis: It is a condition in which two different organisms live together. It is an intimate partnership between two organisms, in which the mutual advantages normally outweigh the disadvantages.

symmetrical: A characteristic of a distribution in which each half is the mirror image of the other half.

sympatric speciation: It refers to the division of one species into two, but in the same geographical locale. Some biologists doubt it occurs.

sympatry: It is the term referring to the relationship between two species or populations when they exist in the same geographic area and thus regularly encounter one another. An initially interbreeding population that splits into two or more distinct species sharing a common range exemplifies sympatric speciation.

symplast: It is the network of highly ordered, connected living axial and radial parenchyma cells in sapwood and inner bark. It is the highly ordered, three-dimensional, connected webwork of living protoplasm in trees.

sympodial branching: It is a type of branching that occurs when the terminal bud ceases to grow (usually because a terminal flower or inflorescence has formed) and an axillary bud or buds take up the further grown the stem.

symptom: It refers to any sensation or change in bodily function experienced by a patient that is associated with a particular disease.

synapsid: It refers to a vertebrate whose skull has one pair of openings behind the eyes.

synapsis: It is the pairing of two homologous chromosomes that occurs during meiosis. It is a form of chromosomal crossover. Synapsis takes place during prophase I. When homologous chromosomes synapse, they come closer together until they are connected by a protein complex called the synaptonemal complex, which contains central and lateral elements.

synaptonemal complex: It is a protein structure that forms between two homologous chromosomes during meiosis and is thought to mediate chromosome pairing, synapsis, and recombination (crossing-over). It is now evident that the synaptonemal complex is not required for genetic recombination.

synchronous (synchronized) culture: It is a microbiological culture or a cell culture that contains cells that are all in the same growth stage.

syncytium: It is a large cell-like structure filled with cytoplasm containing many nuclei. Most cells in eukaryotic organisms have a single nucleus.

syndrome: It is the association of several clinically recognizable features, signs, symptoms, phenomena, or characteristics that often occur together, so that the presence of one feature alerts the physician to the presence of others. In recent decades, the term has been used outside the medicine to refer to a combination of phenomena seen in association.

synergids: These are the cells lying near the micropyle of the embryo sac of an angiosperm.

syngamy: It is the final step in reproduction in which two individuals (gametes) unite permanently and their nuclei fuse. It is a common form of reproduction in protozoa.

synkaryon: It refers to a somatic hybrid cell in which chromosomes from two different parental cells are enveloped in a single nucleus.

syntax: It refer to the rules governing the behaviour of mathematical systems, such as logic, artificial formal languages, and computer programming languages.

synteny: It describes the physical co-localization of genetic loci on the same chromosome within an individual or species. The concept is related to genetic linkage. Linkage between two loci is established by the observation of lower-than-expected recombination frequencies between two or more loci. In contrast, any loci on the same chromosome are by definition syntenic, even if their recombination frequency cannot be distinguished from unlinked loci by practical experiments. Thus, in theory, all linked loci are syntenic, but not all syntenic loci are necessarily linked.

syntocinon: It refers to an injection contains a synthetic version of the naturally-occurring hormone oxytocin. It works in the same way as the natural hormone.

systematic sampling: A systematic sample is selected at random sampling. This method is used when a complete test of the population is available.

T

T cell receptor (TCR): It is a molecule found on the surface of T lymphocytes (or T cells) i.e., in general, responsible for recognizing antigens bound to major histocompatibility complex (MHC) molecules.

T cells or T lymphocyte or lymph cell or lymphocyte: It is an agranulocytic leukocyte that normally makes up a quarter of white blood cell count, but increases in the presence of infection.

T lymphocytes: They are the cells of the immune system belonging to the group of white blood cells known as lymphocytes, and play a central role in cell-mediated immunity.

T: It is the abbreviation for thymine residue in DNA.

T4 DNA ligase: It is a special type of ligase, which is basically an enzyme that in the cell repairs single-stranded discontinuities in double-stranded DNA molecules, in simple words strands that have double-strand break.

T4 DNA Polymerase: It is an enzyme that catalyzes the synthesis of DNA in the $5' \rightarrow 3'$ direction and requires the presence of template and primer. This enzyme has a $3' \rightarrow 5'$ exonuclease activity which is much more active than that found in DNA Polymerase I. Unlike *E. coli* DNA Polymerase I, T4 DNA Polymerase does not have a $5' \rightarrow 3'$ exonuclease function.

tadpole or polliwog or pollywog or polliwig or polwig, or purwiggy or pollywiggle: It is the wholly aquatic larval stage in the life cycle of an amphibian, particularly of a frog or toad.

tag (label): It is an identification slip or note usually tied or written on the experimental sample/specimen.

tag system: It is a deterministic computational model.

tail: It is the section at the rear end of an animal's body.

tailing factor: It is a measure of peak tailing. It is defined as the distance from the front slope of the peak to the back slope divided by twice the distance from the centre line of the peak to the front slope, with all measurements made at 5% of the maximum peak height. The tailing factor of a peak will typically be similar to the asymmetry factor for the same peak, but the two values cannot be directly converted.

tailing: The in vitro addition of the same nucleotide by the enzyme, terminal transferase.

tandem array: It refers to repetitive DNA, where the repeating units are contiguous. Some genes of high copy number occur in tandem arrays.

tandem repeats: It occurs in DNA when a pattern of two or more nucleotides is repeated and the repetitions are directly adjacent to each other.

tandom repeat finder (TRDB): It is a public database of tandem repeats that allows users to run their own sequences.

tank bioreactor: It is type of fermentation bioreactor. A tank bioreactor is a vessel in which a microorganism is grown in a large volume of liquid. This contrasts with fibre or membrane bioreactors and immobilized cell reactors.

tap root: Root system in which the primary root has a much larger diameter than the lateral roots.

tapioca: It is a starch extracted from the root of the plant species *Manihot esculenta.*

***Taq* DNA Polymerase:** It is a thermostable DNA Polymerase isolated from an *E. coli* strain that carries the *Taq* DNA polymerase gene. *Taq* DNA Polymerase is the most common polymerase used for PCR reactions. It is the heat-stable DNA polymerase isolated from the thermophilic bacterium *Thermus aquaticus.*

target site duplication: It is a sequence of DNA that is duplicated when a transposable element is inserted.

target validation: It is a technique which involves proving that DNA, RNA or a protein molecule is directly involved in a disease process and can be a suitable target for development of a new therapeutic drug.

targeted drug delivery: It is a method of delivering medication to a patient in a manner that increases the concentration of the medication in some parts of the body relative to others.

targeting vector: It is a designed cloning vector carrying a DNA sequence capable of targeting a specified chromosomal location in the host cell. Targeting can be used to generate knockouts, knockins or conditional alleles.

target-site overlap (TSO): In a zinc finger protein (ZFP), certain sequences of amino acid residues are able to recognise and bind to an extended target-site of four or even five nucleotides. When this occurs in a ZFP in which the three-nucleotide subsites are contiguous, one zinc finger interferes with the target-site of the zinc finger adjacent to it, a situation called TSO.

TATA box: It is the most common promoter element in eukaryotic protein genes. In eukaryotes, it is a short sequence of base pairs that is rich in adenine (A) and thymidine (T) residues and located about 25-30 nucleotides upstream of the transcriptional initiation site. It is located at –35 to –20. Its consensus sequence, TATAAA, is quite similar to the –10 region of the Sigma 70 recognition site. Another promoter element is called the initiator.

tautomer: It refers to one of two or more structural isomers that exist in equilibrium and are readily converted from one isomeric form to another. Of the various types of tautomerism that are possible, two are commonly observed. In keto-enol tautomerism, a simultaneous shift of electrons and a hydrogen atom occurs; it was first observed by K. Meyer in the ethyl ester of 3-oxobutanoic acid (ethyl acetoacetate), which occurs naturally as a mixture of the two forms.

tautomeric shift: It is the spontaneous isomerization of a nitrogen base to an alternative hydrogen-bonding form, possibly resulting in a mutation. Reversible shifts of proton position in a molecule. Bases in nucleic acids shift between keto and enol forms or between amino and imino forms.

tautomerism: It is the chemical isomerism characterized by relatively easy interconversion of isomeric forms in equilibrium.

taxon: It refers to a classification category for a group of organisms.

T-cell lymphoma: It is a type of lymphoma that affect T cells. This accounts for one in ten cases of non-Hodgkin lymphoma.

T-cell lymphoma: They are the four types of lymphoma that affect T cells. These account for perhaps one in ten cases of non-Hodgkin lymphoma.

T-cell-mediated (cellular) immune response: The immune response produced when sensitized T cells directly attack foreign antigens and secrete lymphokines that initiate the body's humoral immune response.

T-DNA (transfer DNA): It is the DNA of the tumour-inducing (Ti) plasmid of some species of bacteria such as *Agrobacterium tumefaciens* and *Agrobacterium rhizogenes*. It derives its name from the fact that the bacterium transfers this DNA fragment into the host plant's nuclear DNA genome. The T-DNA is bordered by 25-base-pair repeats on each end. Transfer is initiated at the left border and terminated at the right border and requires the *vir* genes of the Ti plasmid.

T-DNA activation tagging: It is a method to generate dominant mutations in plants or plant cells by random insertion of a T-DNA carrying constitutive enhancer elements, which can cause transcriptional activation of flanking plant genes. The method consists of generating a large number of transformed plants or plant cells using a specialized T-DNA construct, followed by selection for the desired phenotype.

technosphere: In ecology, it refers to the part of the physical environment affected through building or modification by humans.

telemeter: It is the unique structure found at the end of eukaryotic chromosomes containing specialized sequences of DNA that assure the completion of a cycle of DNA replication.

telomerase: It is an enzyme that adds DNA sequence repeats ("TTAGGG" in all vertebrates) to the 3' end of DNA strands in the telomere regions, which are found at the ends of eukaryotic chromosomes. This region of repeated nucleotide repeats called telomeres contains condensed DNA material and prevents the constant loss of important DNA from chromosome ends.

telomere: It is a region of repetitive DNA at the end of a chromosome, which protects the end of the chromosome from deterioration. The telomere regions defer the degradation of genes near the ends of chromosomes by allowing for the shortening of chromosome ends, which necessarily occurs during chromosome replication.

telophase: It is a stage in both meiosis and mitosis in a eukaryotic cell. During telophase, the effects of prophase and prometaphase events are reversed. Two daughter nuclei form in the cell. The nuclear envelopes of the daughter cells are formed from the fragments of the nuclear envelope of the parent cell. As the nuclear envelope forms around each pair of chromatids, the nucleoli reappear.

TEM: It is a microscopy technique whereby a beam of electrons is transmitted through an ultra thin specimen, interacting with the specimen as it passes through.

temperate phage: It is a phage that can enter into lysogeny with its host. A phage that can become a prophage.

temperature-sensitive mutant: It is an organism that has a wild-type phenotype at a permissive temperature, but a mutant phenotype at a restrictive (non-permissive) temperature.

temperature-sensitive protein: It is a protein that is functional at one temperature but loses function at another (usually higher temperature).

template strand: DNA (single-stranded) or genomic RNA molecule upon which a complementary nucleotide strand is synthesized. A pattern or mould. DNA stores coded information and acts as a model or template from which information is copied into complementary strands of DNA or transcribed into mRNA. The polymerases use the template to determine the sequence of nucleotides during the synthesis of a new nucleic acid strand.

tentative consensus (TC): It refers to the identification of a sequence from an EST cluster that represents part or all of a complete gene. TCs are usually determined by clustering ESTs allowing for sequencing errors, artefacts such as chimeric clones, and naturally occurring biological phenomena such as alternative splicing. Creation of a cluster allows one to generate a consensus sequence and then identify a long open reading frame which would suggest the possibility of that consensus representing a *bona fide* gene.

Tentative Human Consensus sequences (THCs): It refers to the consensus sequence generated from human EST fragments. THCs may be validated by comparison against databases of known human gene sequences, human genomic sequences, or by identification of the ORFs or other sequence features contained within the consensus as belonging to a known human gene product.

Terminal deoxynucleotidyl transferase (TDT): It is also known as DNA nucleotidylexotransferase (DNTT) or terminal transferase. It is a specialized DNA polymerase expressed in immature, pre-B, pre-T lymphoid cells, and acute lymphoblastic leukemia/lymphoma cells. TdT adds N-nucleotides to the variable, diversity, and joining exons during antibody gene recombination.

termination codons: These are the three codons, UAA known as ochre, UAG as amber and UGA as opal, that do not code for an amino acid but act as signals for the termination of protein synthesis. They are not represented by any tRNA, and termination is catalysed by protein release factors.

termination signal: It is a nucleotide sequence that specifies RNA chain termination.

terminator codons (premature): They are formed by mutation in which one of the three terminator codons in the mRNA used to signal the end of a polypeptide appears in the middle of a genetic message and causes premature termination of transcription and release of incomplete, generally nonfunctional polypeptides from the ribosome.

terminator codons: They are not represented by any tRNA and termination is catalysed by protein release factors. There are two release factors in *E. coli*, RF1 recognises UAA and UAG, RF2 recognises UAA and UGA. Eukaryotes have a single gTP requiring factor, eRF.

terminator: It refers to a DNA sequence just downstream of the coding segment of a gene, which is recognized by RNA polymerase as a signal to stop synthesizing mRNA.

tertiary structure: It refers to the folding of a protein chain via interactions of its sidechain molecules including formation of disulphide bonds between cysteine residues.

test cross: It is a backcross to the recessive parental type and is a cross between an individual exhibiting the dominant phenotype of a trait and an individual that is homozygous recessive for that trait in order to determine the genotype of the dominant individual.

test of independence: A statistical test used to determine whether two or more independent random samples are drawn from the same population or from different populations. In case of independence problems, one sample is taken into consideration, whereas two or more samples are taken from homogeneity.

test of significance: A statistical analysis test used to determine whether observed frequencies between two samples draws from the same population are actually due to the chance or whether they are really significant.

test tube: It is a glass tube in which cells, tissues, etc., are cultured.

test-tube fertilization (in vitro fertilization): It is a method of assisted reproduction—a man's sperm and the woman's egg are combined in a laboratory dish, where fertilization occurs. The resulting embryo is then transferred to the woman's uterus (womb) to implant and develop naturally.

tetracycline: It is a broad-spectrum antibiotic produced by the *Streptomyces* genus of Actinobacteria, indicated for use against many bacterial infections.

tetrad (chromosomes): The group of four chromatids that is formed by the association of duplicated homologous chromosomes during synapsis in meiosis I. They are also known as quadrivalents.

tetrad (pollen): A group of four pollen grains usually found together at the time of formation.

tetraploid: An organism whose cells contain four haploid (4n) sets of chromosomes.

tetrasomics: These are the organisms with four members of one of its chromosomes, whereas the remainder of its chromosome complement is diploid.

thallus: Plant body without true roots, stems, or leaves. It is usually applied to the plants of the group, algae and bryophyte.

therapeutic agent: It refers to a compound that is used for the treatment of a disease or for improving the well-being of an organism.

therapeutic cloning: It is a procedure in which cells, typically skin cells, arc takcn from a patient and inserted into a fertilized egg whose nucleus has been removed. The cell that is so created is permitted to divide repeatedly to form a blastocyst.

therapeutic effect: It is a consequence of a medical treatment of any kind, the results of which are judged to be desirable and beneficial. This is true whether the result was expected, unexpected, or even an unintended consequence of the treatment. An adverse effect, on the other hand, is a harmful and undesired effect.

therapeutic index: It is the ratio between the toxic dose and the therapeutic dose of a drug used.

therapeutic window: It refers to a measure of the relative safety of the drug for a particular treatment. The range of dosage of a drug or its concentration in a bodily system that provides safe effective therapy.

therapeutics: It is a branch that deals specifically with the treatment of disease and the art and science of healing. It also refers to the use of drugs and the method of their administration in the treatment of disease.

thermal pollution: It refers to a temperature rise that threatens an ecosystem, particularly one in which organisms breed or give birth.

thermic shock: It is performed to determine the resistance of the part to sudden changes in temperature. The parts undergo a specified number of cycles, which start at ambient temperature. The parts are then exposed to an extremely low (or high) temperature and, within a short period of time, exposed to an extremely high (or low) temperature, before going back to ambient temperature.

thermolabile: It refers to "not heatproof" and subject to destruction, decomposition, or great change by moderate heating. (used especially for biochemical substances).

thermophile: They are the microbes that reproduce and grow in the temperature range 40° to 70°C (104° to 158°F).

thermosensitivity: It is the central perception of temperature located in the anterior hypothalamus (preoptic region) and spinal cord. The preoptic sensitivity is more important in mammals, the spinal cord centre is more important in birds.

thermostability: It is the quality of a substance to resist irreversible change in its chemical or physical structure at a relative high temperature.

thermotherapy: It refers to a technique in which the plants are exposed to elevated temperatures as a treatment especially to eliminate the contaminating microorganisms. Thermotherapy is used alone or in combination with meristem culture or meristem tip culture.

thymidine block (double): It is a technique used to synchronise the cell cycles of all the cells in a culture. This is done by adding excessive amounts of thymidine (a deoxynucleoside) to the culture so that DNA replication is blocked. Eventually, all of the cells reach the same point in the cell cycle and stop.

thymidine kinase (TK): It is an enzyme, a phosphotransferase (a kinase): 2′-deoxythymidine kinase, ATP-thymidine 5′-phosphotransferase.

thymidine: It is a chemical compound, more precisely a pyrimidine deoxynucleoside. Deoxythymidine is the DNA nucleoside T, which pairs with deoxyadenosine (A) in double-stranded DNA.

thymidylate synthase inhibitors: They are the chemical agents which inhibit the enzyme thymidylate synthase and have potential as an anticancer chemotherapy.

thymidylate synthetase: It is the enzyme used to generate thymidine monophosphate (dTMP), which is subsequently phosphorylated to thymidine triphosphate for use in DNA synthesis and repair.

thymine (T): A pyrimidine base found in DNA. The other organic bases—adenine, cytosine, and guanine—are found in both RNA and DNA.

Ti plasmid: It is a circular plasmid that often, but not always, is a part of the genetic equipment that *Agrobacterium tumefaciens* and *Agrobacterium rhizogenes* use to transduce their genetic material to plants.

tissue culture: It refers to a technique of growing tissues or cells separated from the organism. This is typically facilitated via use of a liquid, semi-solid, or solid growth medium, such as broth or agar. Tissue culture commonly refers to the culture, both of plant and animal cells and is still largely used in the study of developmental phenomena, and mycorrhizal, symbiotic and plant-parasitic relationships.

tissue engineering: It is the use of a combination of cells, engineering and materials methods, and suitable biochemical and physio-chemical factors to improve or replace biological functions.

tonoplast (vacuolar membrane): It is the cytoplasmic membrane that surrounds a vacuole of a plant cell.

topoisomerases: These are the enzymes that unwind and wind DNA, in order for DNA to control the synthesis of proteins, and to facilitate DNA replication. The enzyme is necessary due to inherent problems caused by the DNA's double helix.

totipotency: It refers to the ability of a cell, such as an egg, to give rise to unlike cells and thus to develop into or generate a new organism or part. The property of somatic cells to be induced to undergo regeneration. The diploid zygote formed at fertilization is a single cell which is capable of division and differentiation to give rise to the total range of cell types found in the adult organism. It is the basic principle on the basis of which tissue culture is carried out.

totipotent cell: It is the cell that is capable of developing into any cell type. In a developing embryo, totipotent stem cells are able to produce every cell type in an embryo, including the extraembryonic tissue.

toxicity: It is the degree to which a substance can damage an organism. Toxicity can refer to the effect on a whole organism, such as an animal, bacterium, or plant, as well as the effect on the substructure of the organism, such as a cell (cytotoxicity) or an organ (organotoxicity), such as the liver (hepatotoxicity).

toxin: It is a poisonous substance produced by living cells or organisms.

trace element (microelement): It refers to an element in a sample that has an average concentration of less than 100 parts per million measured in atomic count, or less than 100 micrograms per gram.

tracer: It is an added or injected substance that can be followed within a reaction or an organism, such as radioactive isotopes and certain dyes. A substance used for tracing purposes in histochemistry, the study of the composition of cells and tissues is called histochemical tracer.

transheterozygote: It is used in modern genetics periodicals in two different ways. In the first, the transheterozygote has one mutant (-) and one wildtype allele (+) at each of two different genes (A-/A+ and B-/B+ where A and B are different genes). In the second, the transheterozygote carries two different mutated alleles of the same gene.

transcapsidation: It is the partial or full coating of the nucleic acid of one virus with a coat protein of a differing view.

transcript: A copy of the gene in the form of RNA molecule that has been synthesized from a specific DNA template.

transcription factor: It is a protein that binds to specific DNA sequences, thereby controlling the movement (or transcription) of genetic information from DNA to mRNA.

transcription unit: A stretch of DNA being transcribed into an RNA molecule.

transcription vector: It refers to a cloning vector that allows the foreign gene to be used for generation of transcripts from DNAs which do not have convenient restriction sites at the 3′end.

transcription: It is the process of creating an equivalent RNA copy of a sequence of DNA. Both RNA and DNA are nucleic acids, which use base pairs of nucleotides as a complementary language that can be converted back and forth from DNA to RNA in the presence of correct enzymes.

transcriptional anti-terminator: It is a protein that prevents RNA polymerase from terminating transcription at specific transcription termination sequences. Anti-terminator proteins control gene expression by recognizing control signals near the promoter and preventing transcriptional termination which would otherwise occur at sites that may be a long way downstream.

transduction: It is the process by which DNA is transferred from one bacterium to another by a virus. It also refers to the process whereby foreign DNA is introduced into another cell via a viral vector. This is a common tool used by molecular biologists to stably introduce a foreign gene into a host cell's genome.

transfection: It is the process of deliberately introducing nucleic acids into cells.

transfer RNA (tRNA): It is a small RNA molecule (usually about 74-95 nucleotides) that transfers a specific active amino acid to a growing polypeptide chain at the ribosomal site of protein synthesis during translation.

transferase: It is an enzyme that catalyzes the transfer of a functional group (e.g., a methyl or phosphate group) from one molecule (donor) to another (acceptor).

transferred DNA (T-DNA): It is the transferred DNA of the tumour-inducing (Ti) plasmid of some bacteria. It is about 20,000 base pairs long and contains genes that code for enzymes synthesizing opines and phytohormones. By transferring the T-DNA into the plant genome, the bacterium essentially reprograms the plant cells to grow into a tumour and produce a unique food source for the bacteria.

transformant: It refers to a cell that has been genetically altered through the uptake of foreign DNA. In higher eukaryotes, a cultured cell that has acquired a malignant phenotype.

transformation efficiency: It is the efficiency by which cells can take up extracellular DNA and express genes encoded by it. This is based on the competency of the cells. It can be calculated by dividing the number of successful transformants by the amount of DNA used during a transformation procedure.

transformation frequency: It is the rate at which fraction of a cell population takes up foreign DNA. It is expressed as the number of transformed cells divided by the total number of cells in a population.

transformation: It is the genetic alteration of a cell resulting from the uptake, incorporation and expression of exogenous genetic material (DNA) that is taken up through the cell wall.

transforming oncogene: A gene that upon transfection converts a previously immortalized cell to the malignant phenotype.

transgene: It is a segment of DNA containing a gene sequence that has been isolated from one organism and is introduced into a different organism. This non-native segment of DNA may retain the ability to produce RNA or protein in the transgenic organism, or it may alter the normal function of the transgenic organism's genetic code. In general, the DNA is incorporated into the organism's germ line.

transgenesis: It is the process of introducing an exogenous gene, called a transgene, into a living organism so that the organism will exhibit a new property and transmit that property to its offspring.

transgressive variation: The production of offspring with phenotypes that exceed the parental extremes.

transhumanism: It is an international intellectual and cultural movement supporting the use of science and technology to improve human mental and physical characteristics and capacities.

transient: It refers to short duration.

transition: It refers to the substitution in DNA or RNA of one purine by another purine, or one pyrimidine by another pyrimidine.

translation: It is the third stage of protein biosynthesis (part of the overall process of gene expression). In translation, messenger RNA (mRNA) produced by transcription is decoded by the ribosome to produce a specific amino acid chain, or polypeptide, that will later fold into an active protein. Translation occurs in the cell's cytoplasm, where the large and small subunits of the ribosome are located, and bind to the mRNA. The ribosome facilitates decoding by inducing the binding of tRNAs with complementary anticodon sequences to that of the mRNA. The tRNAs carry specific amino acids that are chained together into a polypeptide as the mRNA passes through and is "read" by the ribosome in a fashion reminiscent to that of a stock ticker and ticker tape.

translocation: It is a chromosome abnormality caused by rearrangement of parts between nonhomologous chromosomes. A gene fusion may be created when the translocation joins two otherwise separated genes, the occurrence of which is common in cancer.

transmembrane region: It refers to the region of a transmembrane protein that actually spans the membrane. Transmembrane regions are usually hydrophobic in order to be thermodynamically compatible with the lipid bilayer portion of the membrane. They may consist of either alpha-helical or beta-strand secondary structure elements, but in either case the external residues (the ones facing the membrane) are invariably hydrophobic while the internal residues may be hydrophilic (as in the case of a pore or channel) or polar. One common transmembrane structural domain is the seven-helix bundle seen in numerous channel proteins.

transmission electron microscope (TEM): It is a microscope that operates on the same basic principles as the light microscope but uses electrons instead of light. What you can see with a light microscope is limited by the wavelength of light. TEMs use electrons as "light source", and their much lower wavelength makes it possible to get a resolution a thousand times better than with a light microscope.

transposable genetic element: It is a DNA element that can move from one location in the genome to another.

transposase: It is an enzyme encoded by a transposon gene that catalyses the movement of a DNA sequencc to a different site in a DNA molecule, by catalysing the excision of the transposon from one site and its insertion into a new chromosomal site.

transposons: They are also called transposable elements or jumping genes, or transposons. They are a heterogeneous class of genetic elements that can insert at new locations on chromosomes. Transposons are classified into three major groups: Class I, retrotransposons; Class II, DNA transposons; Class III, miniature inverted-repeat transposable elements. They have direct repeat flanking sequences that are not part of the transposons, and terminal inverted repeats that are part of the transposons.

transversion: It refers to the substitution of a purine for a pyrimidine or vice versa. It can only be reverted by a spontaneous reversion. Because this type of mutation changes the chemical structure dramatically, the consequences of this change tend to be more severe and less common than that of transitions.Transversions can be caused by ionizing radiation and alkylating agents.

treatment: The object of comparison in the experimental trials.

tribrid protein: It is a fusion protein that has three segments, each encoded by parts of different genes.

trinucleotide repeat disorders: They are also known as trinucleotide repeat expansion disorders, triplet repeat expansion disorders or codon reiteration disorders. These are a set of genetic disorders caused by trinucleotide repeat expansion, a kind of mutation where trinucleotide repeats in certain genes exceed the normal, stable, threshold, which differs per gene. The mutation is a subset of unstable microsatellite repeats that occur throughout all genomic sequences. If the repeat is present in a healthy gene, a dynamic mutation may increase the repeat count and result in a defective gene.

triplet codon: Codon is a sequence of three nucleotides which codes for a specific amino acid. For example, the codon ATG/AUG codes for methionine. When three sets of amino acid code for a particular character, they are called triplet codon.

triploid syndrome: It is an extremely rare chromosomal disorder. Individuals with triploid syndrome have three of every chromosome for a total of sixty-nine rather than the normal forty-six chromosomes. Babies with Triploid Syndrome are usually lost through early miscarriage. However, some infants have been born and survived as long as five months. Affected infants are usually small and have multiple birth defects. Those that survive are usually mosaic, meaning that some cells have the normal number of 46 chromosomes and some cells have a complete extra set of chromosomes.

triploid: It refers to an organism containing three times the haploid number of chromosomes.

trisomy: It is a genetic abnormality in which there are three copies, instead of the normal two, of a particular chromosome.

triticale: It is a man-made crop developed by crossing wheat (*Triticum turgidum* or *Triticum aestivum*) with rye (*Secale cereale*).

tRNA (transfer RNA): It is a small RNA molecule (usually about 74-95 nucleotides) that transfers a specific active amino acid to a growing polypeptide chain at the ribosomal site of protein synthesis during translation. It has a 3' terminal site for amino acid attachment. This covalent linkage is catalyzed by an aminoacyl tRNA synthetase. It also contains a three base region called the anticodon that can base pair to the corresponding three base codon region on mRNA.

true-to-type: It denotes correct cultivar identification and lack of variation in productivity or performance. Verification is done visually by an expert or through biochemical, serological or other means.

trypsin: It is a serine protease found in the digestive system of many vertebrates, where it hydrolyses proteins. It is produced in the pancreas as inactive proenzyme trypsinogen. It cleaves peptide chains mainly at the carboxyl side of the amino acids lysine or arginine, except when either is followed by proline. It is used for numerous biotechnological processes.

tubulin: They are the major protein components of the microtubules of eukaryotic cells. Microtubules are obligate proteinaceous elements found in nearly all eukaryotic cells.

tumble tube: It is a glass tube mainly used in vitro to agitate and consequently aerate suspension cultures.

tumour virus: It refers to a virus that is capable of transforming a cell to a malignant

tundra: It refers to an arctic or subarctic treeless plain with frozen subsoil and small growth: some grasses, dwarf shrubs, lichens, mosses, sedges, and herbs. Often found at high latitudes between the tree line and the snow line.

turbidostat: It is a continuous culture device, similar to a chemostat or an auxostat, which has feedback between the turbidity of the culture vessel and the dilution rate.

turgid: It means swollen and distended or congested. It usually refer to a cell that is firm due to water uptake.

turgor pressure: It is the main pressure of the cell contents against the cell wall in plant cells and bacteria cells, determined by the water content of the vacuole, resulting from osmotic pressure, i.e., the hydrostatic pressure produced by a solution in a space divided by a semipermeable membrane due to a differential in the concentration of solute.

type I error: Rejecting a null hypothesis when it is true.

type II error: Accepting a null hypothesis when it is false.

U

U: The uracil residue in RNA.

ubiquitin ligase (E3 ubiquitin ligase): It is a protein that in combination with an E2 ubiquitin-conjugating enzyme causes the attachment of ubiquitin to a lysine on a target protein via an isopeptide bond; the E3 ubiquitin ligase targets specific protein substrates for degradation by the proteasome. In general, the ubiquitin ligase is involved in polyubiquitination: A second ubiquitin is attached to the first, a third is attached to the second, and so forth.

ubiquitin: It is a small regulatory protein that has been found in almost all the cells with nuclei (eukaryotes). It directs proteins to recycling and other functions. Ubiquitin binds to proteins and labels them for destruction. The ubiquitin tag directs proteins to the proteasome, which is an organelle in the cell that degrades and recycles the proteins that are not required.

ubiquitous (omnipresent): Being or seeming to be everywhere at the same time.

Ultrasonic: It is concerned with, or producing waves with the same nature as sound waves but frequencies above audio frequencies.

ultrasonography: It refers to the imaging of deep structures of the body by recording the echoes of pulses of ultrasonic waves directed into the tissues and reflected by tissue planes where there is a change in density. Diagnostic ultrasonography uses 1–10 megahertz waves.

ultrasound: It is cyclic sound pressure with a frequency greater than the upper limit of human hearing. Although this limit varies from person to person, it is approximately 20 kilohertz (20,000 hertz) in healthy, young adults and thus, 20 kHz serves as a useful lower limit in describing ultrasound. The production of ultrasound is used in many different fields, typically to penetrate a medium and measure the reflection signature or supply focused energy.

ultrastructure: It is the detailed structure of a biological specimen, such as a cell, tissue, or organ, that can be observed by electron microscopy. It refers, in general, to the study of cellular structures that are too small to be seen with an optical microscope. Ultrastructure, along with molecular phylogeny, has often been a reliable (that is, phylogenetic) way of classifying organisms.

ultraviolet light: It is electromagnetic radiation with a wavelength shorter than that of visible light, but longer than X-rays, in the range 10 nm to 400 nm, and energies from 3eV to 124 eV. It is so named because the spectrum consists of electromagnetic waves with frequencies higher than those that humans identify as the colour violet.

unbiased error: Unbiased errors arise due to chance differences between the members of the population included in the sample and those not included in the sample.

understock: It refers to the host plant for a grafted scion, a branch or shoot from another plant.

understock: It refers to the segment of plant to which a graft is joined.

unequal crossing over: It is the crossing over between homologous chromosomes that are not precisely paired, resulting in nonreciprocal exchange of material and chromosomes of unequal length. Favoured in regions containing tandemly repeated sequences. A genetic defect occurring in crossing over where one of the homologous chromosomes receives more genetic information from the other chromosome than it passes over, therefore meaning that one of them has duplicated information, while the other lacks.

Unidentified reading frame (URF): It refers to the open reading frame that is encoding a protein of undefined function.

uniformitarianism: It refers to the well-substantiated hypothesis that the earth and its organisms developed primarily from gradual processes and conditions.

unisexual: It refers to the animals and plants possessing either male or female reproductive organs, but not both.

univalent: It is relating to a chromosome that is not paired or united with its homologous chromosome during synapsis.

univariate analysis: Statistical analysis in which only one variable is involved.

universal donor cells: Cells that, after introduction into a recipient, will not induce an immune response that leads to their rejection.

universal donor: A person who is type O in the ABO blood group system and can donate blood to all recipients.

upper tailed-test: A one tailed hypothesis test in which a sample value significantly above the hypothesised population value will lead us to reject the null hypothesis.

upstream: It refer to a relative position in DNA or RNA. Each strand of DNA or RNA has a 5' end and a 3' end, so named for the carbons on the deoxyribose (or ribose) ring. Relative to the position on the strand, downstream is the region towards the 3' end of the strand. Since DNA strands run in opposite directions, downstream on one strand is upstream on the other strand.

uracil: It is a common and naturally occurring pyrimidine derivative.

V

vaccination schedule: It is a series of vaccinations, including the timing of all doses, which may be either recommended or compulsory, depending on the country of residence.

vaccination: It is the administration of antigenic material (a vaccine) to produce immunity to a disease. Vaccines can prevent or ameliorate the effects of infection by many pathogens. There is strong evidence for the efficacy of many vaccines, such as the influenza vaccine.

vaccine: It is a biological preparation that improves immunity to a particular disease. A vaccine typically contains an agent that resembles a disease-causing microorganism, and is often made from weakened or killed forms of the microbe or its toxins. The agent stimulates the body's immune system to recognize the agent as foreign, destroy it, and "recognize" it, so that the immune system can more easily recognize and destroy any of these microorganisms that it later encounters.

vacuole: It is a membrane bound organelle which is present in all plant and fungal cells and some protist, animal and bacterial cells. Vacuoles are essentially enclosed compartments which are filled with water containing inorganic and organic molecules including enzymes in solution, though in certain cases they may contain solids which have been engulfed.

vacuolization: It is the formation of vacuoles within or adjacent to the cells, and, in dermatopathology, often refers to the basal cell-basement membrane zone area.

vacuum gauge: It is a measure used to calculate the pressure in a vacuum, which is further divided into two subcategories: high and low vacuum (and sometimes ultra-high vacuum). The applicable pressure range of many of the techniques used to measure vacuums have an overlap. Hence, by combining several different types of gauge, it is possible to measure system pressure continuously from 10 mbar down to 10^{-11} mbar.

vacuum pump: It is a device that removes gas molecules from a sealed volume in order to leave behind a partial vacuum. The vacuum pump was invented in 1650 by Otto von Guericke.

vacuum: It is a volume of space that is essentially empty of matter, such that its gaseous pressure is much less than the atmospheric pressure.

Variable bitrate (VBR): It is a term used in telecommunications and computing that relates to the bitrate used in sound or video encoding. As opposed to constant bitrate (CBR), VBR files vary the amount of output data per time segment. VBR allows a higher bitrate (and therefore more storage space) to be allocated to the more complex segments of media files, while less space is allocated to less complex segments. The average of these rates can be calculated to produce an average bitrate for the file.

variable domains: These are the regions of antibodies showing variations. The general structure of all antibodies is very similar. A small region at the tip of the protein is extremely variable, allowing millions of antibodies with slightly different tip structures, or antigen binding sites, to exist.

variable expressivity: It refers to a phenomenon of variation in the phenotype caused by different alleles of the same gene.

Variable numbers of tandem repeats (VNTRs): It refers to the DNA sequence blocks of 2-60 base pairs which are repeated from two to more than 20 times in different individuals. This polymorphism makes VNTRs very useful DNA markers used in genomic mapping, linkage analysis and also DNA fingerprinting.

variable surface glycoprotein (VSG): One of a battery of antigenic determinants expressed by a microorganism to elude immune detection.

variable: A quantity or quality liable to show variation from one individual to the next in the same population.

variance: It is defined as the mean of squares of deviations.

variant: It refers to an organism having or exhibiting variation or differing from the other.

variate: An individual observation of any variable.

variegated: It refers to plant pigmentation that creates multicoloured foliage.

variegated: Plants having discrete markings of different colours on leaves.

variety: It refers to a naturally occurring subdivision of a species, with distinct morphological characters.

vascular: It refers to the plant tissue or region consisting of or giving rise to conducting tissue usually made up of xylem and phloem. The plants having this tissue are called vascular plants.

vector: It refers to an entity or an organism that carries or transmits a molecule or a disease causing organism.

vegetative propagation: Asexual or non-sexual method of reproduction.

vehicle: Referring to host organism used for the replication or expression of a

velocity density gradient centrifugation: It is a method or procedure used to separate macromolecules based on their rate of movement through a density gradient.

velogenetics: It is a method in which marker-assisted selection and embryo technologies are used together in order to increase the rate of genetic improvement in animal populations.

vermicompost: It is the excreta of earthworm, which is rich in humus. Earthworms eat cow dung or farmyard manure along with other farm wastes and pass it through their body, in the process converting it into vermicompost.

vermiculite: It is a natural mineral that expands with the application of heat. The expansion process is called exfoliation and is routinely accomplished in purpose-designed commercial furnaces. Vermiculite is formed by weathering or hydrothermal alteration of biotite or phlogopite.

vermiculture: It refers to the rearing or cultivation of earthworms, and the technology is the scientific process of using them for the betterment of human beings.

vernalization: It is the acquisition of a plant's ability to flower or germinate in the spring by exposure to prolonged cold of winter.

vessel: It refers to a series of xylem elements whose function is to conduct water and nutrients in plants.

viability of cells: The status of a cell to survive, grow, and multiply.

viability test: It is an assay or a test to determine the proportion of living individuals, cells or organisms in a sample. Viability tests are most commonly performed on cultured cells, and usually depend on the ability of living cells to exclude a dye, (an exclusion test) or specifically take it up (inclusion test).

viable: A status referring to be able to take root and grow.

vibrio: Comma-shaped bacterium, e.g., *Vibrio comma,* water-borne bacterium which causes asiatic cholera.

***vir* genes:** It is a set of genes on a Ti plasmid that prepare the T-DNA segment for transfer into a plant cell. They are induced by signals produced by plants following wounding. Phenolic compounds such as acetosyringone activate the *VirA* gene, which is a constitutively expressed transmembrane protein. The activated *VirA* gene acts as a kinase, phosphorylating the *VirG* gene. In its phosphorylated form *VirG* acts as an activator of transcription for the remaining Vir gene operons. Vir genes are necessary for the transfer of T-DNA, but they function *in trans*, so none of these genes need to be included in the cloning vectors. Modified *Agrobacterium* strains provide all the necessary Vir functions on modified Ti plasmids.

viral oncogene: It is a gene that contributes to malignancies in vertebrate hosts.

viral proteins: They are the proteins generated by a viruses. Many are structural, forming the viral envelope and capsid. However, there are also viral nonstructural proteins and viral regulatory and accessory proteins.

viral vaccines: Vaccines consisting of live viruses rather than dead ones or separated parts of viruses. These live virus vaccines are prepared from attenuated strains that are almost or completely devoid of pathogenicity but are capable of inducing a protective immune response. They multiply in the human host and provide continuous antigenic stimulation over a period of time. Primary vaccine failures are uncommon and are usually the result of inadequate storage or administration.

virion: It is a complete virus particle with its DNA or RNA core and protein coat as it exists outside the cell. It is also called a viral particle.

viroid: They are plant pathogens that consist of a short stretch (a few hundred nucleobases) of highly complementary, circular, single-stranded RNA without the protein coat that is typical for viruses. In comparison, the genome of the smallest known viruses capable of causing an infection by themselves are around 2 kilobases in size. The human pathogen, hepatitis D, is now known to be caused by viroids. Viroids were discovered by Theodor Otto Diener.

virtual libraries: It refers to the creation and storage of vast collections of molecular structures in an electronic database. These databases may be queried for subsets that exhibit specific physicochemical features, or may be "virtually screened" for their ability to bind a drug target. This process may be performed prior to the synthesis and testing of the molecules themselves.

virtual water: It is the hidden flow of water if food or other commodities are traded from one place to another. The precise volume of water can be more or less depending on climatic conditions and agricultural practice.

virulence factors: They are molecules expressed and secreted by pathogens (bacteria, viruses, fungi and protozoa) that enable them to achieve colonisation of a niche in the host, evasion of the host's immune response, inhibition of the host's immune response, entry into and exit out of cells, and obtaining nutrition from the host.

virulence: It is the degree of pathogenicity within a group or species of microorganisms or viruses as indicated by case fatality rates and/or the ability of the organism to invade the tissues of the host. The pathogenic capacity of an organism is determined by its virulence factors.

virulent phage: It refers to a phage (virus) that always leads to lysis of the host cell.

virus: It is a protein-coated fragment of DNA or RNA that infects a host cell in order to reproduce.

virus-free: An organism which exhibits no viral symptoms or contains no identifiable virus-particles.

virus-tested: It is a stock of things or organisms certified as being free of certain specified viruses following recognized procedures of virus diagnosis.

visible spectrum: It is the portion of the electromagnetic spectrum that is visible to (can be detected by) the human eye. Electromagnetic radiation in this range of wavelengths is called visible light.

visualization: It refers to the process of representing abstract scientific data as images that can aid in understanding the meaning of the data.

vitamins: They are organic compounds required as a nutrient in tiny amounts by an organism. They cannot be synthesized in sufficient quantities by organisms, and must be obtained from the diet.

vitrified (water soaked): It refers to cultured tissue having leaves and sometimes stems with a glassy, transparent or wet and often swollen appearance.

vitrify: To change or make into glass or a glassy substance, especially through heat fusion.

vivipary: It is a form of reproduction in animals in which the developing embryo obtains its nourishment directly from the mother via a placenta or by other means.

V_{max}: It refer to the maximal rate of an enzyme-catalysed reaction.

VNTR (Variable number tandem repeat): It is a location in a genome where a short nucleotide sequence is organized as a tandem repeat. These can be found on many chromosomes, and often show variations in length between individuals. Each variant acts as an inherited allele, allowing them to be used for personal or parental identification. Their analysis is useful in genetics and biology research, forensics, and DNA fingerprinting.

volatilization (vaporization): The conversion of a chemical substance from a liquid or solid state to a gaseous or vapour state by the application of heat, by reducing pressure, or by a combination of these processes.

volatilize: To become or make volatile.

volume (V): The amount of space occupied by a three-dimensional object or region of space, expressed in cubic units.

volume/volume (V/V): When the solute is a liquid, sometimes it is convenient to express its concentration in volume/volume percent (v/v %).

VOQ: An abbreviation for Virtual Output Queuing. It is a strategy in which each input port maintains a separate queue for each output port.

W

wall of cell: It is the tough, usually flexible, but sometimes fairly rigid layer that surrounds some types of cells. It is located outside the cell membrane and provides these cells with structural support and protection, and also acts as a filtering mechanism. A major function of the cell wall is to act as a pressure vessel, preventing over-expansion when water enters the cell. They are found in plants, bacteria, fungi, algae, and some archaea. Animals and protozoa do not have cell walls.

wall pressure: It is the mechanical pressure exerted by the cell contents against the cell wall. The rigidity of a plant often depends on the level of wall pressure found in the cells of the stem. Wall pressure falls if the plant cell loses water.

Wallace effect: It refers to the females attracted to males most like themselves, breed offspring more like themselves and less like the members of any subspecies; a tendency that eventually creates a separate species.

wash-out: The loss of the slower growing microorganism when two organisms are being grown together.

water crisis: It refers to the world's water resources relative to human demand. The term has been applied to the worldwide water situation by the United Nations and other world organizations.

water cycle (hydrologic cycle or H_2O cycle): It is continuous movement of water on, above and below the surface of the earth. Water can change states among liquid, vapour, and ice at various places in the water cycle.

water pollution: It is the undesirable change in the physical, chemical and biological nature of water. When toxic substances enter lakes, streams, rivers, oceans, and other water bodies, they get dissolved or lie suspended in water or get deposited on the bed. This results in the pollution of water, whereby the quality of the water deteriorates, affecting aquatic ecosystems. Pollutants can also seep down and affect the groundwater deposits.

water potential: It is the potential energy of water per unit volume relative to pure water in reference conditions. Water potential quantifies the tendency of water to move from one area to another due to osmosis, gravity, mechanical pressure, or matrix effects such as surface tension. Water potential has proved especially useful in understanding water movement within plants, animals, and soil. Water potential is typically expressed in potential energy per unit volume and very often is represented by the Greek letter Ø.

water stress: It occurs when the demand for water exceeds the available amount during a certain period or when poor quality restricts its use. Scarcity can be absolute, such as in environments of low precipitation and large transpiration rates. It can also, however, be induced by economic or political constraints, which do not permit the adequate development of water resources.

water table: It refers to the pooling of groundwater on top of a bedrock layer. This pooling holds 95% of the world's total supply of fresh water.

water: It is a chemical substance with the chemical formula H_2O. Its molecule contains one oxygen and two hydrogen atoms connected by covalent bonds. Water is a liquid at ambient conditions, but it often co-exists on the earth with its solid state, ice, and gaseous state, water vapour or steam.

wavelength: It is the spatial period of the wave—the distance over which the wave's shape repeats. It is usually determined by considering the distance between consecutive corresponding points of the same phase, such as crests, troughs, or zero crossings, and is a characteristic of both travelling waves and standing waves, as well as other spatial wave patterns.

wax: It refers to a class of hydrocarbons that are plastic (malleable) at normal ambient temperatures.

waxoyl: It is a rust-preventative fluid used to prolong the life of metallic instruments by excluding water and air from bare metal and by slowing the progress of existing surface rust using phosphoric acid rust inhibitors.

weathering: It is the breaking down of the earth's rocks, soils and minerals through direct contact with the planet's atmosphere. Weathering occurs in situ, or "with no movement", and thus should not be confused with erosion, which involves the movement of rocks and minerals by agents such as water, ice, wind, and gravity.

weed: It is a plant that is considered by the user of the term to be a nuisance, and normally applied to unwanted plants in man-made settings.

weight matrix: Density of binding sites in a gene or sequence can be used to derive a ratio of density for each element in a pattern of interest. The combined individual density ratios of all elements are then collectively used to build a scoring profile known as a weight matrix. This profile can be used to test the prediction of the identification of the selected pattern and the ability of the algorithm to discriminate them from non-pattern sequences.

weight per volume (w/v): It is representation of the relationship of a solute to a solvent expressed as grams of solute per millilitre of the total solution. An example is 50 g of glucose in 1 L of solution, considered a 5% w/v solution.

western blot (protein immunoblot): It is an analytical technique used to detect specific proteins in a given sample of tissue homogenate or extract. It uses gel electrophoresis to separate native or denatured proteins by the length of the polypeptide or by the 3-D structure of the protein. The proteins are then transferred to a membrane, where they are probed using antibodies specific to the target protein.

wet weight: The living material containing the chemical of interest is not dried to remove water. The amount of chemical found in subsequent analysis is expressed as the weight of chemical divided by the total weight, including any water present, of the material which once contained it.

wetting agents: They are the chemical substances that increase the spreading and penetrating properties of a liquid by lowering its surface tension, that is, the tendency of its molecules to adhere to each other.

wetting rate (minimum): It is the smallest liquid-flow rate through a packed column that will thoroughly wet the column packing.

white biotechnology: It is the application of biotechnology for industrial purposes, including manufacturing, alternative energy (or "bioenergy"), and biomaterials. It includes the practice of using cells or components of cells like enzymes to generate industrially useful products.

wild type: It refers to the phenotype of the typical form of a species as it occurs in nature.

wilting point: A condition in which a plant begins to use water from its own tissues for transpiration because soil water has been exhausted.

wilting: It refers to a plant disease characterized by drooping and shriveling; usually caused by parasites attacking the roots.

within-sample variance: An estimate of the population variance based on the variance within the 'k' samples, using a weighted average of the 'k' sample variance.

wobble hypothesis: It is the hypothesis proposed by Crick in 1966. According to the wobble hypothesis, only the first two positions of a triplet codon on mRNA have a precise pairing with the bases of the tRNA anticodon. The triplet code is a degenerate one with many more codons than the number of amino acid types coded.

wobble: The ability of certain bases at the third position of an anticodon in tRNA to form hydrogen bonds in various ways, causing alignment with several possible codons. Referring to the reduced constraint of the third base of an anticodon as compared with the other bases, thus allowing additional complementary base pairings.

X

xanthophyll: A yellow chloroplast pigment.

x-chromosome: A chromosome associated with sex determination.

x-chromosome: It is one of the two sex-determining chromosomes in many animal species, including mammals (the other is the Y chromosome). It is a part of the XY sex-determination system and X0 sex-determination system.

xenia: The immediate effect of pollen on some characters of the endosperm.

xenobiotic: It is a chemical which is found in an organism, but which is not normally produced or expected to be present in it. It can also cover substances which are present in much higher concentrations than are usual. Specifically, drugs such as antibiotics are xenobiotics in humans because the human body does not produce them itself, nor are they part of a normal diet.

xenoglossy (xenoglossia): It is the putative paranormal phenomenon in which a person is able to speak or write a language he or she could not have acquired by natural means.

xenograft: It refers to a surgical graft of tissue from one species to an unlike species.

xenon: It is a chemical element represented by the symbol Xe.

xenophobia: It is an irrational, deep-rooted fear of or antipathy towards foreigners.

xenopus: It is a genus of highly aquatic frogs native to Sub-Saharan Africa.

xenotransplantation: It is the transplantation of living cells, tissues or organs from one species to another, such as from pigs to humans.

xeric: It is a condition characterized by, or adapted to an extremely dry habitat.

xerophytes: They are the plants which are able to survive in an environment with little available water or moisture, usually in environments where potential evapotranspiration exceeds precipitation for all or part of the growing season.

X-linked disease: It is a genetic disease caused by an allele at a locus on the X-chromosome. Since males have only one X chromosome, any mutated gene on the X- chromosome will result in disease.

X-linked inheritance: Transfer of characters from parents to offspring is inheritance, and the inheritance through x-chromosomes is called x-linked inheritance. The trait determined by a gene on the X chromosome is always expressed in males.

x-ray crystallography: It is a method of determining the arrangement of atoms within a crystal, in which a beam of X-rays strikes a crystal and diffracts into many specific directions. From the angles and intensities of these diffracted beams, a crystallographer can produce a three-dimensional picture of the density of electrons within the crystal. From this electron density, the mean positions of the atoms in the crystal can be determined, as well as their chemical bonds, disorder, and various other information.

x-ray scattering techniques: They are a family of non-destructive analytical techniques which reveal information about the crystallographic structure, chemical composition, and physical properties of materials and thin films.

x-ray tube: It is a vacuum tube that produces X-rays.

x-ray: It is a form of electromagnetic radiation. X-rays have a wavelength in the range of 0.01 to 10 nanometers, corresponding to frequencies in the range 30 petahertz to 30 exahertz and energies in the range 120 eV to 120 keV.

xylem vessels: They are a long straight chain made of tough long dead cells known as vessel elements. The cells are arranged end to end, and the cell walls which would normally surround the whole cell dissolve whereever they are touching another cell wall.

xylem: It is a tissue specialized for efficient conduction of water and mineral nutrients in solution.

Y

y-axis: It is the line on a graph that runs vertically (up-down) through zero. It is used as a reference line from which one can measure from it.

y-chromosome: It is one of the two sex-determining chromosomes in most mammals, including humans. In mammals, it contains the gene SRY, which triggers testis development if present. The human Y chromosome is composed of about 60 million base pairs.

Yeast (*Saccharomyces cerevisiae*): It is commonly known as baker's or brewer's yeast, and has been used in research for a very long time. The complete genome sequence of the *S. cerevisiae* lab strain S288C was obtained in the spring of 1996, making yeast the first eukaryotic organism to be completely sequenced. The complete sequence of its genome has proven extremely useful as a reference toward extracting the sequences of human and other higher eukaryotic genes. Furthermore, the ease of genetic manipulation of yeast allows its use for conveniently analyzing and functionally dissecting gene products from other eukaryotes.

Yeast 2-hybrid system: It refers to the yeast-based method used to simultaneously identify and clone the gene for proteins interacting with a known protein. The basis of this method is a "transcriptional reporter assay" (see definition) in which reporter gene expression is dependent on two domains. The first domain is linked to the known protein. The second domain is genetically linked to a library. If the library is screened against the known protein, the two domains will interact only if a protein from the library binds the known protein, resulting in transcription activation of the reporter gene, and a blue colour.

yeast artificial chromosome (YAC): It is a vector used to clone DNA fragments larger than 100 kb and up to 3000 kb and can accommodate foreign DNA inserts of 200 to 500 kb. YACs are useful for the physical mapping of complex genomes and for the cloning of large genes. It was first described in 1983 by Murray and Szostak. It contains the telomeric, centromeric, and replication origin sequences needed for replication and preservation in yeast cells.

yeast episomal vector (YEp): These are the vectors that carry 2 mm circles of DNA sequence including the origin of replication and rep genes. One example is YEp13. This yeast cloning vector is a shuttle vector having a selectable LEU 2 gene.

yeast extract: It is the common name for various forms of processed yeast products made by extracting the cell contents (removing the cell walls); they are used as food additives or flavourings, or as nutrients for bacterial culture media.

yeast: It is an unicellular, eukaryotic fungus, commonly found as a contaminant in plant tissue culture. Dandruff is a minor infection of the scalp caused by an overabundance of a yeast-like fungus that is part of the normal skin flora.

yield (crop): It is a measure of produce of crop per unit area of land under cultivation; yield is also the seed generation of the plant itself.

Y-intercept: A constant for any given straight line whose value represents the value of the Y variable when the X variable has a value of O.

Z

Z-DNA (zig-zag DNA): It is one of the many possible double helical structures of DNA. It is a left-handed double helical structure in which the double helix winds to the left in a zig-zag pattern. It was the first crystal structure of a DNA molecule that has been managed to be solved by scientists. At present no definite biological significance is associated with the presence of Z-DNA. But it is commonly believed that it gives a torsional strain relief while DNA transcription is happening.

Z-DNA: It refers to the conformation of DNA existing as a left-handed double helix (the phosphate-sugar backbone forms a left-handed zig-zag course), which may play a role in gene regulation.

Zebrafish (*Danio rerio*): It is used as a model organism because of its small size, short life cycle, ease of culture, and ability to readily produce mutations relevant to human health and disease. The embryonic development of the zebrafish can be seen through its transparent egg and closely resembles that of higher vertebrates, making it useful for studying development and mutations. Other shared features with humans include blood, kidney, and optical systems. In addition, its genome is half the size of the mouse and human genomes, which is valuable in identification of key vertebrate genes.

zebrafish: It is a tropical freshwater fish and its zoological name is *Danio rerio*. It is a common and useful model organism for the study of vertebrate development and gene function. Fully-sequenced genetic code. It is useful in research because it is well understood, easily observable and testable developmental behaviours and well-characterized mutants are available.

Zero Population Growth (ZPG): It is the growth when birth rate (plus immigration) equals the death rate (plus emigration), resulting in an ecologically stable number of organisms.

zinc fingers: It refers to the protein motif formed by the interaction of repeated cysteine and histidine residues with a zinc ion. The spacing of the repeats results in finger-like arrangements of the protein loops formed from the interaction which interact with DNA. These motifs are typically found in transcription factors.

zone of elongation: It refers to a region in the young root or shoot just behind the apical meristem, in which the cells are enlarging and elongating rapidly.

zone of inhibition: It refers to a clear region; is an indication of the absence, or the effective inhibition of microbial growth by the antimicrobial agent.

zoo blot (garden blot): It is a type of Southern blot that demonstrates the similarity between specific, usually protein-coding, DNA sequences of different species. A zoo blot compares animal species, while a garden blot compares plant species.

zoo FISH: It refers to a fluorescent in situ hybridization of DNA from one species on metaphase chromosomes of another species.

zoonotic diseases: It refers to diseases transmissible from animals to humans. Ebola, tuberculosis, and rabies are examples.

zooplankton: It refers to plankton that do not produce their own food, as phytoplankton do. In size they fall between phytoplankton and larger organisms like fish.

zoospermia: The presence of live spermatozoa in the ejaculated semen.

zoospore: It is a motile asexual spore that uses a flagellum for locomotion. Also called a swarm spore, these spores are created by some algae, bacteria and fungi to propagate themselves.

zygomatic bone (cheekbone or malar bone): It is a paired bone of the human skull. It articulates with the maxilla, the temporal bone, the sphenoid bone and the frontal bone.

zygote: It is diploid cell synthesized from the union of two gametes, and constitutes the first stage in a unique organism's development.

zymogen (proenzyme): It is an inactive enzyme precursor.

ZymoGenetics: It is an American pharmaceutical company involved in the development of therapeutic proteins.

zymogenous: The term refers to microorganisms, often transient or alien, that respond rapidly by enzyme production and growth when simple organic food sources become available.

Appendix 1: The Amino Acids, Symbols, and Codons

Amino acids	*Three letter symbols*	*One letter symbols*	*Codons*
Alanine	Ala	A	GCA, GCC, GCG, GCU
Cysteine	Cys	C	UGC, UGU
Aspartic acid	Asp	D	GAC, GAU
Glutamic acid	Glu	E	GAA, GAG
Phenylalanine	Phe	F	UUC, UUU
Glycine	Gly	G	GGA, GGC, GGG, GGU
Histidine	His	H	CAC, CAU
Isoleucine	Ile	I	AUA, AUC, AUU
Lysine	Lys	K	AAA, AAG
Leucine	Leu	L	UUA, UUG, CUA, CUC, CUG, CUU
Methionine	Met	M	AUG
Asparagine	Asn	N	AAC, AAU
Proline	Pro	P	CCA, CCC, CCG, CCU
Glutamine	Gln	Q	CAA, CAG
Arginine	Arg	R	AGA, AGG, CGA, CGC, CGG, CGU
Serine	Ser	S	AGC, AGU, UCA, UCC, UCG, UCU
Threonine	Thr	T	ACA, ACC, ACG, ACU
Valine	Val	V	GUA, GUC, GUG, GUU
Tryptophan	Trp	W	UGG
Tyrosine	Tyr	Y	UAC UAU

Appendix 2: Genetic Code Dictionary

	T	*C*	*A*	*G*
T	TTT Phe (F)	TCT Ser (S)	TAT Tyr (Y)	TGT Cys (C)
	TTC Phe (F)	TCC Ser (S)	TAC	TGC
	TTA Leu (L)	TCA Ser (S)	TAA STOP	TGA STOP
	TTG Leu (L)	TCG Ser (S)	TAG STOP	TGG Trp (W)
C	CTT Leu (L)	CCT Pro (P)	CAT His (H)	CGT Arg (R)
	CTC Leu (L)	CCC Pro (P)	CAC His (H)	CGC Arg (R)
	CTA Leu (L)	CCA Pro (P)	CAA Gln (Q)	CGA Arg (R)
	CTG Leu (L)	CCG Pro (P)	CAG Gln (Q)	CGG Arg (R)
A	ATT Ile (I)	ACT Thr (T)	AAT Asn (N)	AGT Ser (S)
	ATC Ile (I)	ACC Thr (T)	AAC Asn (N)	AGC Ser (S)
	ATA Ile (I)	ACA Thr (T)	AAA Lys (K)	AGA Arg (R)
	ATG Met (M) START	ACG Thr (T)	AAG Lys (K)	AGG Arg (R)
G	GTT Val (V)	GCT Ala (A)	GAT Asp (D)	GGT Gly (G)
	GTC Val (V)	GCC Ala (A)	GAC Asp (D)	GGC Gly (G)
	GTA Val (V)	GCA Ala (A)	GAA Glu (E)	GGA Gly (G)
	GTG Val (V)	GCG Ala (A)	GAG Glu (E)	GGG Gly (G)

Appendix 3: Common Genetic Disorders in Man

Genetic disorder	*Mutation*	*Chromosome*
22q11.2 deletion syndrome	D	22q
Angelman syndrome	DCP	15
Colour blindness	P	X
Cri du chat	D	5
Cystic fibrosis	P	7q
Down syndrome	C	21
Duchenne muscular dystrophy	D	Xp
Haemophilia	P	X
Klinefelter's syndrome	C	X
Neurofibromatosis	NF 1	17q/22q
Phenylketonuria	P	12q
Prader-Willi syndrome	DC	15
Sickle-cell disease	P	11p
Tay-Sachs disease	P	15
Turner syndrome	C	X

Appendix 4: Nitrogenous Bases and Nucleosides

Nitrogenous base	*Nucleoside*	*Deoxynucleoside*
Adenine	Adenosine A	Deoxyadenosine dA
Guanine	Guanosine G	Deoxyguanosine dG
Thymine	5′-Methyluridine m^5U	Thymidine dT
Uracil	Uridine U	Deoxyuridine dU
Cytosine	Cytidien C	Deoxycytidine dC

Appendix 5: Measurements of DNA

bp = base pair(s) one bp (= 3.4 Å)

kb (= kbp) = kilo base pairs = 1,000 bp

Mb = mega base pairs = 1,000,000 bp

Gb = giga base pairs = 1,000,000,000 bp.

Tb = tera base pairs = 1,000,000,000,000 bp.

Eb = exa base pairs = 1,000,000,000,000,000 bp

Zb = zetta base pairs = 1,000,000,000,000,000,000 bp.

Yb = yotta base pairs = 1,000,000,000,000,000,000,000 bp.

Bb = bronto base pairs = 1,000,000,000,000,000,000,000,000 bp.

Appendix 6: Table of Bytes and Computer Memory

1 bit = a single digit, either 1 or 0

8 bits = 1 byte, a combination of 1's and 0's

1024 Bytes = 1 KB (kilobyte)

1024 Kilobytes = 1 MB (megabyte)

1024 Megabytes = 1 GB (gigabyte)

1024 Gigabytes = 1 TB (Terabyte)

1024 Terabytes = 1 PB (Petabyte)

1024 Exabytes = 1 Zettabyte

1024 Zettabytes = 1 Yottabyte

1024 Yottabytes = 1 Brontobyte

1024 Brontobytes = 1 Geopbyte

Appendix 7: Greek Alphabets

A	α	Alpha
N	ν	Nu
B	β	Beta
Ξ	ξ	Xi
Γ	γ	Gamma
O	ο	Omicron
Δ	δ	Delta
Π	π	Pi
E	ε	Epsilon
P	ρ	Rho
Z	ζ	Zeta
Σ	σς	Sigma
H	η	Eta
T	τ	Tau
Θ	θ	Theta
Y	υ	Upsilon
I	ι	Iota
Φ	φ	Phi
K	κ	Kappa
X	χ	Chi
Λ	λ	Lambda
Ψ	ψ	Psi
M	μ	Mu
Ω	ω	Omega

Appendix 8: Metric Prefixes Used in Molecular Biology

m	(mili)	10^{-3}	k	(kilo)	10^{3}
micro (μ)	(micro)	10^{-6}	M	(mega)	10^{6}
n	(nano)	10^{-9}	G	(giga)	10^{9}
p	(pico)	10^{-12}	T	(tera)	10^{12}
f	(femto)	10^{-15}	P	(peta)	10^{15}
a	(atto)	10^{-18}	E	(exa)	10^{18}
z	(zepto)	10^{-21}	Z	(zeta)	10^{21}
			Y	(yotta)	10^{24}